Climate Change, Vulnerability and Migration

Edited by
S. Irudaya Rajan
and R. B. Bhagat

Routledge
Taylor & Francis Group

LONDON AND NEW YORK

First published 2018 by Routledge

2 Park Square, Milton Park, Abingdon, Oxfordshire OX14 4RN

52 Vanderbilt Avenue, New York, NY 10017

Routledge is an imprint of the Taylor & Francis Group, an informa business

First issued in paperback 2019

British Library Cataloguing-in-Publication Data
A catalogue record for this book is available from the British Library

Library of Congress Cataloging-in-Publication Data
A catalog record has been requested for this book

ISBN: 978-0-415-79072-7 (hbk)
ISBN: 978-0-367-34541-9 (pbk)

Typeset in Sabon
by Apex CoVantage, LLC

Climate Change, Vulnerability and Migration

This book highlights how climate change has affected migration in the Indian subcontinent. Drawing on field research, it argues that extreme weather events such as floods, droughts, cyclones, cloudbursts as well as sea level rise, desertification and declining crop productivity have shown higher frequency in recent times and have depleted biophysical diversity and the capacity of the ecosystem to provide food and livelihood security. This volume shows how the socio-economically poor are worst affected in these circumstances and resort to migration to survive.

The chapters in this volume study the role of remittances sent by migrants to their families in environmentally fragile zones in providing an important cushion and adaptation capabilities to cope with extreme weather events. This book looks at the socio-economic and political drivers of migration, different forms of mobility, mortality and morbidity levels in the affected population, and discusses mitigation and adaption strategies.

This volume will be of great interest to scholars and researchers of environment and ecology, migration and diaspora studies, development studies, sociology and social anthropology, governance and public policy and politics.

S. Irudaya Rajan is Professor, Centre for Development Studies, Thiruvananthapuram, Kerala, India. With more than three decades of research experience, he has coordinated with K. C. Zachariah seven major migration surveys in Kerala since 1998, conducted migration surveys in Goa (2008) and Tamil Nadu (2015) and provided technical support to Gujarat (2010) and Punjab (2011) migration surveys. He is editor of the annual series 'India Migration Report' and the editor-in-chief of the journal *Migration and Development*.

R. B. Bhagat is Professor and Head, Department of Migration and Urban Studies, International Institute for Population Sciences (IIPS), Mumbai, Maharashtra, India. He has served as Consultant to the UNESCO-UNICEF India Initiative on Migration and to the International Organisation of Migration (IOM) and Advisor to the Yale University Project on Climate Change and Communication. His research interests are in population, urbanisation, environment and migration issues.

Dedicated to our teachers who inspired
us for migration research

PROFESSOR K C ZACHARIAH
LATE PROFESSOR S MUKERJI

Contents

Illustrations

Figures

Tables

Maps

Boxes

Contributors

Soumyadeep Banerjee works with the International Centre for Integrated Mountain Development, Kathmandu, Nepal, and University of Sussex, Brighton, United Kingdom.

R. B. Bhagat is Professor and Head, Department of Migration and Urban Studies, International Institute for Population Sciences (IIPS), Mumbai, Maharashtra, India.

Mohammad Rashed Alam Bhuiyan is Assistant Professor, Department of Political Science, University of Dhaka, Bangladesh.

Richard Black is the Pro-Director (Research & Enterprise) at the SOAS University of London, United Kingdom.

Abdul Jaleel C. P. is a research scholar at the International Institute for Population Sciences, Mumbai, India.

Shareena Banu C. P. is Assistant Professor, Department of Sociology, Jamia Millia Islamia, New Delhi, India.

Aparajita Chattopadhyay is Assistant Professor, Department of Development Studies, International Institute for Population Sciences, Mumbai, India.

Bhaswati Das is Associate Professor, Centre for the Study of Regional Development, Jawaharlal Nehru University, New Delhi, India.

Partha Jyoti Das is Head of the Water Climate and Hazard (WATCH) Programme, Aaranyak, Guwahati, India.

Bratati Dey is Assistant Professor, Department of Geography, Dhruba Chand Halder College, West Bengal, India.

Giovanna Gioli works with the International Centre for Integrated Mountain Development (ICIMOD), Kathmandu, Nepal.

Md. Towheedul Islam is Assistant Professor, Department of International Relations, University of Dhaka, Bangladesh, and a research scholar, Department of Political and Social Change, The Australian National University, Australia.

Bhagwati Joshi is Assistant Professor of Geography, Government Post Graduate College, Rudrapur, Uttarakhand, India.

Dominic Kniveton is Professor of Climate Science and Society, Department of Geography, School of Global Studies, University of Sussex, Brighton, United Kingdom.

Vijay Korra is Assistant Professor, Centre for Economic and Social Studies, Hyderabad, India.

Ritumbra Manuvie is Commonwealth Scholar, School of Politics and International Relations, University of Edinburgh, United Kingdom.

Maxmilan Martin is a teaching fellow in Interdisciplinary Environmental Anthropology/Human Ecology at University College London, United Kingdom.

Avijit Mistri is Assistant Professor, Department of Geography, Nistarini Women's College, Sidho-Kanho-Birsha University, West Bengal, India.

Divya Mohan is a Science Policy Officer with the Indian Himalayas Climate Adaptation Programme (IHCAP) Programme Management Unit (PMU), New Delhi.

Architesh Panda works with the International Rice Research Institute, Los Baños, Philippines.

S. Irudaya Rajan is Professor, Centre for Development Studies, Thiruvananthapuram, Kerala, India.

Tasneem Siddiqui is Professor, Department of Political Science and Founding Chair, Refugee and Migratory Movements Research Unit (RMMRU), University of Dhaka, Bangladesh.

Prakash C. Tiwari is Professor of Geography, Kumaon University, Nainital, India.

Himani Upadhyay works at the Disaster Research Unit (DRU), Freie University Berlin and The Energy and Resources Institute (TERI), New Delhi.

Preface and acknowledgements

Migration has acquired a centrestage in the areas of sustainable development, politics and gender. Several studies show that the environmental and climate change has been shaping the magnitude and intensity of migration, albeit interwoven with a number of socio-economic and political drivers. It is now evident that a large number of people are getting affected by the rising temperature and changes in the pattern of precipitation. The extreme weather events like floods, droughts, cyclone, cloudbursts and more frequently threaten agriculture and food security in many parts of the world including India. This may also raise mortality and morbidity levels along with migration in the affected population if mitigation and adaption strategies are not followed.

India is a vast country and some parts or the other always experience floods, cyclones and droughts. The frequency and intensity of such events have increased in recent times, which are a plausible sign of climate change. This book largely focuses on the multifaceted dimensions of migration and issues pertaining to climate change in India.

This book illuminates on climate change vulnerabilities to migration based on papers presented in two international conferences organised under the aegis of UNESCO, New Delhi, and Centre for Development Studies (CDS), Thiruvananthapuram, Kerala, respectively. This volume carefully selected some of the papers presented in the 'National Workshop on Migration and Global Environmental Change in India', 4–5 March 2014, organised by UNESCO and Foresight, United Kingdom, under the guidance of Ms Marina Faetanini, Chief, Social and Human Sciences (SHS), UNESCO, New Delhi. We thankfully acknowledge her support and permission to include some of the papers presented in the workshop. Also, a few papers were selected from an international seminar on 'Migration, Care Economy and Development', organised at CDS in honour of Professor K. C. Zachariah. In addition, we also invited papers from known scholars working in the areas of climate change and its impact on migration and livelihood strategies.

This manuscript contains 13 articles and an introduction highlighting how extreme weather events like floods and droughts affect migration. It argues that these extreme events which show higher frequencies in recent times depletes biophysical diversity and the capacity of the eco-system to provide food and livelihood security. The socio-economically poor are worst affected, and many resort to migration to survive. On the other hand, remittances sent by them to the families left behind in the vulnerable and affected areas provide important cushion and adaptation to cope with climate change–induced extreme weather events.

We sincerely thank the contributors to this volume who have accomplished an excellent mission in unravelling the intricate and complex relationship between migration and climate change.

1 Migration in the context of climate change

An introduction

R. B. Bhagat and S. Irudaya Rajan

The context

There is hardly any dispute among scientists that the climate change has emerged as one of the most devastating threats to the mankind. In order to study the nature and consequences of climate change the United Nations Environment Programme (UNEP) and the World Meteorological Organization (WMO) established an Intergovernmental Panel on Climate Change (IPCC), which was endorsed by the United Nations General Assembly in 1988. Climate change is recognized by IPCC as a significant man-made global environmental challenge and published its first assessment report in 1990, following which the UN adopted the United Nations Framework Convention on Climate Change in 1992. It is worthwhile to reiterate that from its first report the IPCC warned about the impacts of climate change on human migration (Gomez 2013). The IPCC, which submitted its fifth assessment report recently, continued to provide increasing evidences on climate change. There is a growing consensus among climate scientists that global warming is taking place, precipitations are erratic and there is an increase in extreme weather events like floods, droughts and cyclones globally (Trenberth et al. 2007; Stern 2007; INCCA 2010). The Paris Agreement on climate change adopted in 2016 was a most recent global response to the threat of climate change. It envisages to keep the global temperature rise this century below 2°C, above pre-industrial levels or even to limit further increase to 1.5°C. To achieve this, countries have committed to implement their Nationally Determined Contributions (NDCs). India is committed to reduce its emissions intensity of GDP by 33–35 per cent by 2030 from its 2005 level and protect its population from the adverse impacts of climate change. Further, countries have flexibility under the agreement to determine NDCs, according to their national circumstances.

As far as the migration is concerned, the Paris Agreement recognizes the impact of climate change on vulnerable populations and reaffirms to protect their rights. Migrant population was recognized as one of the vulnerable populations along with indigenous, local children and populations with disabilities (http://unfccc.int/files/essential_background/convention/application/pdf/english_paris_agreement.pdf; accessed on 15 October 2016). Similarly, out of the 17 Sustainable Development Goals (SDGs) that the world community had agreed in September 2015 to pursue by 2030, three goals mention migration and migrants dealing with the SDGs. Goal 8 'Economic Growth and Decent Work' recommends to protect labour rights and promote safe and secure working environments for all workers, including migrant workers, in particular women migrants, and those in precarious employment. Goal 10 'Reduce Inequality within and among Countries' suggests to facilitate orderly, safe, regular and responsible migration and mobility of people, including through the implementation of planned and well-managed migration policies and goal 17 in respect with 'Global Partnership' acknowledges data gaps in the field of migration and recommends to strengthen disaggregated data including migratory status.

Climate change manifests in two forms, namely gradual and sudden. Gradual changes are increasing aridity and desertification, loss of biodiversity, rising sea level and so on, while sudden changes include extreme weather events like cyclone, floods, severe droughts and more. A number of researchers have studied the impact of climate change on migration. Lester Brown introduced the term 'environmental refugees' in the early 1970s, which was later defined by El-Hinnawi (1985) as people displaced through either natural disaster or gradual environmental degradation. Some researchers estimated the figure of environmental refugees to rise as high as 150 million by the year 2050 (Myers 2002; United Nations University 2007). However, the concept of environmental refugees was criticized for its vagueness by several scholars who suggested the term 'environmental migrant' as an alternative (Black 1998; Bates 2002). In fact, climate change acts as a push factor intertwined with economic issues (Adamo 2009). According to Myers (2002: 610), 'poverty serves as an additional "push" factor associated with the environmental problems that displace people. Other factors include population pressures, malnutrition, landlessness, unemployment, over-rapid urbanization, pandemic diseases and government shortcomings, together with ethnic strife and conventional conflicts. In particular, it is sometimes difficult to differentiate between refugees that are driven by environmental factors and those that are impelled by economic problems'. For many people in India and Sub-Saharan

Africa, poverty combined with lack of environmental resources has been the main motivating forces for migration. For these people, it matters little whether they view themselves primarily as environmental or economic refugees (Myers 2002: 610).

In the study on migration, the classic conceptualization of push and pull factors provided the initial but the most potent basis of explanations about why people migrate. Push factors are related to the areas of origin which are mainly dependent on agriculture and allied economic activities, while pull factors are associated with areas of destination having potential of industrial expansion and increasing opportunities of jobs in the service sectors (Heberle 1938). Further, in addition to economic factors, push and pull factors also encompass a number of social and political factors such as ethnic and social conflict in the areas of origin, and political inclusiveness and integration of migrant community with that of the host in the areas of destination affecting the nature and magnitude of migration. In the trade-off between push and pull factors, some researchers have assumed the precedence of push factors operating as drivers of migration and considered the environmental and climate change as one of the drivers of migration, as a consequence of the depleting sources of livelihood and threat to the survival of people at the place of origin.

Historically viewed, the environmental change and natural disaster have always been a push factor of migration. However, since the late 20th century, climate change has emerged as an important determinant of environmental change and natural disasters shaping migration and mobility of people (Laczko and Aghazaram 2009). In analysing the impact of climate change, the frame of differentiated responsibility but differential vulnerability is important from policy perspectives and global action. The developed nations and the richer households have contributed largely to the emission of greenhouse gases in the past, while the poor, indigenous and marginal communities in the less developed nations have to bear the brunt due to loss of livelihood, food insecurity and risk of health, disease and displacement. In most international conventions and agreements, this aspect has featured as a core issue and efforts are made to achieve environmental justice and protection of rights. Studies show that the impact of climate change is likely to be more severe in poor countries with high population growth (Samson et al. 2011). Many areas in Asia and Africa have been experiencing high internal and international outmigration also (UNDP 2009; Foresight 2011). Challenges for some of the large countries with huge number of poor like India, Pakistan, Bangladesh, Nigeria and Ethiopia will be daunting. Thus, the regional vulnerability to the climate change is an important issue and we have tried to address some of the concerns in this volume.

Climate change as a driver of migration

Experts working in the areas of environmental change perhaps first realized the disastrous effects of weather changes in the 1970s and 1980s in the form of forced migration, namely displacement, raising the attention of the international community about the seriousness of the problem (Gomez 2013). Since then, several studies on climate change and migration have thrown light on the issue (Raleigh et al. 2008; Bardsley and Hugo 2010; Piguet et al. 2011; Gray and Bilsborrow 2013), but the subject still requires continued and extensive attention as climate change operates at the global level while effects are felt locally.

Climate change is likely to affect migration through a number of socio-economic and political drivers and its direct effect is confounded by the overlapping nature of climate change and socio-economic conditions. The environmental factors as the result of climate change are interwoven with economic, social and political factors. In many cases, environmental drivers manifest through economic drivers. In such a situation, it is not always easy to identify the exact contribution of environmental and climate change as a cause of migration. However, it must be emphasized that it would be wrong to attribute migration as a monocausal phenomenon (Mazumdar et al. 2013), rather it would be appropriate to attribute migration resulting from a multi-causality of the interwoven and embedded nature of socio-economic, political and environmental factors expressed through a livelihood strategy. Climate change adds to the economic stress pushing people to seek livelihood outside their area of residence or are displaced in the event of climate change. Although the precise outcome of migration as a result of climate change is difficult to predict, it is however possible to know the number of people affected by climate changes due to rising temperature and uncertain precipitation, often seen in extreme weather events like floods, droughts, cyclone and cloudbursts. Sea level rise, desertification and declining crop productivity are other processes affecting the livelihood of the people, forcing many of them to migrate. It is also worthwhile to emphasize that different migration-causing factors act at different spatial and temporal scales that make the outcome difficult to generalize, yet migration due to livelihood deterioration is an ongoing process and environmental migrants seem to be an increasing challenge. There are various forms of migration. Both internal and international migrations are likely to be affected by climate change. Evidences show that climate change may lead to internal than international migration, but internal migration is likely to predominate (Massey et al. 2010; Bardsley and Hugo 2010). However, within internal migration, the seasonal and temporary migration could exacerbate with deteriorating natural resources compelling people to adopt migration as a livelihood strategy. Such type of migration outcome is likely to be slow, interwoven with a number of other socio-economic factors; researchers call this a linear migration, mostly taking place through the established routes and network of migration. On the other hand, non-linear migration would be sudden and huge and likely to adopt paths and destinations unknown or less known in the trajectory of migration. Most of this migration will take the form of displacement (Bardsley and Hugo 2010).

Migration as an adaption to climate change

Climate change increases risk and vulnerability of those people who have been experiencing it. Risks denote the potential of shocks and stresses that affect the state of systems, communities, households and individuals. On the other hand, the vulnerability is the propensity or predisposition to be adversely affected. It is a dynamic concept that varies over space and time and depends on economic, social, geographic, demographic, cultural, institutional, governance and environmental factors (IPCC 2012). The risks and vulnerability must have been assessed on the backdrop of adaptive capacity. It is a dynamic concept shaped by the interaction of environmental, social, cultural, political and economic forces that determine vulnerability through exposures and sensitivities, and the way the system's components are internally reacting to shocks. In fact, it has two dimensions: adaptive capacity to absorb shocks (coping ability) and adaptive capacity to change over time (adaptability, management capacity; Smit and Wandel 2006).

The populations affected by adverse climate change frequently employ *in situ* (in place) adaptation strategies as a first response towards livelihood insecurities (Bardsley and Hugo 2010; Davis and Lopez-Carr 2010). There are numerous examples of in situ adaptation strategies which include sales of assets, intensifying livelihood activities or adopting new ones, use of formal and informal credit, reducing non-essential expenditures and drawing on social networks and public programmes for assistance (Gray and Mueller 2012). If those strategies prove to be insufficient or infeasible, a household might decide to send a member elsewhere or, as an option of last resort, the entire household might decide to relocate (Warner 2011). Migration could be a possible adaptive mechanism but several studies also show that migration is socio-economically selective; this means that not everybody affected by climate change would be privileged to migrate and escape its wrath. However, the potential role of migration and remittances sent by migrants may play a very important role in mitigating human distress arising in the event of climate change. A growing number of research findings demonstrate that migration can have a transformative power, remittances could promote adaption strategies and distress reduction as people adopt circular mobility patterns and livelihood strategies based on links to multiple locations (IOM 2015: 119). It helps transfer surplus labour from agriculture to non-agriculture sector, increasing the efficiency of labour use and enhancing productivity and reduction in poverty. Further, researchers have also pointed out that migration has been an informal process of skill development. Migrants learn new skills and knowledge from co-workers and friends at the place of destination (Bhagat 2014). During the time of climate change–induced disasters, remittances could support the families left behind and also motivate them to adopt agricultural and economic activities in such a manner that might lead to adaptation in the changed circumstances. In addition to remittances, migrant groups and associations can support various development and risk reduction projects in areas of origin such as food and water security, infrastructural provisions and technical support along with political mobilization in support of the distressed population (IOM 2015: 99).

Climate change also raises a crucial concern about how resources, technologies and regulations are to be used to support the victims of climate change in the coming decades. At destination, migrant communities have been disproportionately affected by loss of income and livelihoods related to the physical destruction of homes and displacement as a result of climate change–induced extreme weather events and disasters. Planned and well-managed migration could be more helpful instead of leaving to the people to resort to unplanned migration as an adaptive strategy (IOM 2015). Thus, policies are required to reap the benefits of migration and reduce its negative consequences. The adaptive capacity of migration and consequent remittances need to be leveraged with other ongoing programmes at the place of origin to increase migration efficiency and its development, while there should be exclusive policies for the protection against the vulnerability of migrants at the place of destination which are mostly the urban areas. The impact of urban floods, cyclones, landslides and other climate change disasters are very obvious in urban areas in different parts of the world. The migrants are one of the most vulnerable groups that need to be protected through suitable urban policy and planning. It is a good practice that United Nations has proclaimed 18 December as International Migrants Day to highlight the plight of the migrants across the world. It is up to the state to formulate suitable policies and programmes not only to protect the human rights of the migrants and their families, but also to devise means to integrate them in the development process and planning.

Although climate change has been a global phenomenon, the impact is likely to vary across countries, regions and also overtime. As various studies across the world show that vulnerability is the key issue in the context of climate change and adaption is a coping strategy, this book focuses on these issues largely in the context of India. India is not only the second most populous country in the world with a population of 1.3 billion, but also a country with the largest number of poor (363 million; Rangarajan and Dev 2014). Various studies show that the poor are more vulnerable to climate change. As the push factors appear to be very strong, the role of environmental and climate change cannot be ignored in India's migration history. We also find that there is a dearth of work on internal migration in South Asia in general and India in particular in the context of climate change, and also the forms of internal migration such as rural-to-urban migration or short-term seasonal temporary migration shaped by climate change. This manuscript fills these gaps with a focus on India and its neighbours, namely Pakistan and Bangladesh.

Organization of the volume

The manuscript contains 13 articles highlighting how extreme weather events like floods and droughts, and also sea level rise, affect migration, remittances and adaptation. It argues that these extreme events, which show higher frequencies in recent times, deplete biophysical diversity and capacity of the ecosystem to provide food and livelihood security. The socio-economically poor are worst affected, and many resort to migration to survive. On the other hand, remittances sent by them to the families left behind in the vulnerable and affected areas provide important cushion and adaptation to cope with climate change–induced extreme weather events. In the chapter 'Climate change vulnerability and migration in India: overlapping hotspots', R. B. Bhagat highlights that India is one of the most vulnerable countries to climate change as a large area falls under vulnerable regions of coastal zones, mountains and dry lands. It also shows that the vulnerability in these regions was compounded by underdevelopment and a large population dependent on rain-fed agriculture. This chapter also presents a theoretical framework that underpinned on identifying overlapping hotspots for migration and environmental change. The concept of vulnerability is central to the theoretical framework, and is defined as the likelihood that an individual or group is exposed to the adverse impact of climate change and their capacity to adjust to it. The theoretical framework consists of a series of variables – exposure to climate change; the underlying socioeconomic conditions; external drivers such as globalization, political factors and conflicts – that either singularly or in combination influence the vulnerability of communities leading to mobility outcomes as a livelihood strategy. The framework viewed vulnerability as the net outcome of interaction between climate change stress and the socioeconomic capacity to adapt. Migration is one of the potential means of adaption to climate change. The study examined the pattern of migration in relation to the exposure of Indian agriculture to the impacts of climate change and globalization. On the other hand, coastal states of India are more urbanized and some of them are net inmigrating states. It is estimated that a population of about 73 million live in the low elevation coastal zone (the area located between 0 and 10 metres above sea level) comprising about 3 per cent of the land area and 6 per cent of India's population. The study demonstrated that high inmigration and high biophysical vulnerability overlap in the case of coastal regions. In dry land regions that are characterized by high biophysical vulnerability, the matrix indicated that there is a very high outmigration as well. Thus, it is essential to take into account the migration outcomes of climate change, its role in promoting adaption and resilience in the areas of origin. In this line of thought, Upadhyay and Mohan, in the chapter 'Migrating to adapt? Exploring the climate change, migration and adaptation nexus', have questioned the dominant narrative and the discourses on migration highlighting the need for *in situ* adaptation and cited cases of Maldives and Lakshadweep islands. They opined that popular discourses on migration as adaptation or as failure to adapt do not resonate with on-the-ground perspectives of people in regions threatened by sea level rise. The authors argue that migration as an adaptation to climate change can be constrained by variations in risk perceptions, the distant nature of climate change and failure to link current experiences with future events. In respect to the islands, a sense of 'place belongingness' is important for people in terms of their identity, culture, local traditions and community cohesion. Differing from the catastrophic framings of the issue of climate change on small islands, they added that climate change is not an everyday priority in the region. However, it does not mean that the problem does not exist but important to enhance local research capacities, and migration as an adaptation option should be explored locally. The chapter concludes by stating that there is a need to promote *in situ* adaptation focusing on local priorities, people's choices and their institutional contexts rather than simply to foster migration.

Changes in the frequency and intensity of floods and droughts are significant manifestations of climate change. It could be a product of either climate variability (yearly basis) or climate change (long-term changes in the average weather conditions) often to be determined by climate scientists. For social scientists, consequences on the lives of the people and their struggle to cope with the wrath of extreme weather events like floods and droughts are of utmost importance. Banerjee and colleagues, in their chapter 'Exploring vulnerability in flood-affected remittance-recipient and non-recipient households of upper Assam in India', emphasize a growing consensus among migration scholars that remittances tend to be a counter-cyclical shock absorber in times of crisis. However, the extent to which remittances can contribute to climate change adaptation among remittance-recipient households is complex and requires further exploration. They point out that previous research has adopted an index-based approach to examine vulnerability of a country, community, sector or ecosystem. However, similar methodology has seldom been applied to explore whether remittances have a role in reducing the vulnerability of recipient households to a particular environmental stressor. This chapter presents an empirical study of floods in the Eastern Brahmaputra sub-basin. Authors found that village-level flood preparedness remains low, and household-level flood responses are mainly comprised of short-term and reactive strategies. The remittances are crucial to meet the basic needs (e.g. food, education, healthcare) of remittance-recipient households. The findings from the vulnerability assessment indicate that differences between remittance-recipient and non-recipient households are significant primarily at the attribute level. When these attributes are aggregated into sub-dimensions, and in turn the sub-dimensions are aggregated into major components, these differences between two groups of household tend to disappear. It is likely that different attributes cancel each other upon aggregation at the next higher level in hierarchy. However, an insight about attributes of household-level sensitivity and adaptive capacity is no less useful for the local government institutions, development partners and community-based organizations from the perspective of local adaptation planning. This would help to design specific interventions for the households. For example, non-farm income diversification is an attribute of reduction on environmental dependence. Local government institutions could organise non-farm skill training opportunities for the youth and women. This would help to diversify the household portfolio, and in turn minimise the risk from extreme events. In another study from the state of Assam, Manuvie evaluates the institutional response given to communities displaced due to chronic flood disasters. It is argued that both academics and non-academic commentators rely on the principles of economic and political rationality to analyse the desirability of the strategies adopted to offset the annual damage caused due to floods. In the process, cost-effectiveness of the flood mitigation and flood management policies are scrutinized to provide an assessment of the policy and implementation gaps. The current chapter, however, questions this ontology of criticism about governmental failure by arguing that the laggard process of policy building and response towards disasters and disaster-induced displacements is normatively desirable to achieve the required level of functionality in a divided society. While there is a strong recognition of the fact that the gradual development of the policy response is far from ideal and jeopardizes the essential rights of disaster victims, the chapter extends the paradigm of incrementalism and argues that a muddled through response is rather essential and humane method in which day-to-day administration can be managed in a deeply divided society. This chapter draws on analyses and interpretations from the high-level elite interviews with bureaucrats in the State of Assam and highlights how social and political constructions in a society necessitate incremental responses instead of drastic policy changes.

In a similar but somewhat a different study, Gioli analysed the role of migration and remittances in mitigating the adverse impact of flood and conflict in North-Western Pakistan. In the chapter 'Remittances as self-insured life: on migration, flood and conflict in North-Western Pakistan', the author points out that prominent actors such as the International Organisation for Migration (IOM) increasingly promote remittances and migration as a way of supporting the resilience of populations in the context of natural or man-made catastrophes and crises. The rationale behind this idea is that both conflict and large-scale natural disasters occur in context where informal work, precarious land rights and subsistence agriculture, along with the lack of access to financial instruments and other forms of social protection severely limit the ability of people to cope with crisis and insure themselves against risks. However, in contrast to this argument, the author presents an alternative narrative from a field work carried out in 2012 in the Swat and Lower Dir districts of Khyber-Paktunkhwa,

Pakistan. The study documents the perceived role of remittances in the wake of both natural shocks (the 2010 flood) and conflict (2009 Taliban insurgency). Remittances proved to be a crucial means of survival for the affected population, and labour migration acts as a form of self-insurance strategy against shocks. However, without the right policy, excessive emphasis on migration as self-help mechanism might actually lead to perpetuating both local and global disparities and vulnerabilities rather than help overcoming them. On the other hand, Siddiqui and colleagues, in the chapter 'Situating migration in planned and autonomous adaptation practices to climate change in Bangladesh', highlight the policy discourse of Bangladesh viewing migration as a negative outcome of climate change. Such understanding led almost all the actors, that is, the Government of Bangladesh, its international development partners and also the non-governmental organizations, to concentrate on infrastructure interventions and local-level livelihood adaptation programmes to reduce the need for migration. This paper demonstrates that contrary to planned adaptation programmes, the autonomous adaptation practices of affected households have integrated different forms of migration as one of the many adaptation strategies. This paper used both primary and secondary data from three regions, all hotspots of environmental and climate change hazards. Chapai Nawabganj suffers from drought, Satkhira experiences cyclones and saline intrusion and Munshiganj faces floods and riverbank erosion, which are also common in the previous two areas. It establishes that adaptation outcome of migration is context specific and under some social, economic, demographic and policy environment migration that leads to adaptation while in some other situation it leads to maladaptation.

It is well known that sea level rise, apart from islands, also threatens some of low-lying delta areas. The Sundarban Delta spreading over India and Bangladesh is one such coastal area threatened by the sea level rise. Mistri and Das show in the chapter 'Migration in response to environmental change: a risk perception study from Sundarban Biosphere Reserve' that over the years, environmental migration has been a growing concern in the Sundarbans. Outmigration is a prominent livelihood strategy in the Sundarban. At least one member from three-fourth of the households migrated from the area in search of work. This chapter presents an insight into the subjective appraisal of the environmental change in the Sundarban, its influence on means of livelihood, especially on farming and fishing, and finally its linkage to outmigration. It is a comparative study between two groups, migrants and non-migrants. Both the migrants and non-migrants perceive more or less equal environmental risk, and the impact of environmental change is subsumed under socio-economic conditions. Climate could be considered as a hidden driver of migration and working in tandem with other predominant factors. The factors like assets, capabilities, structure and process (i.e. governmental and non-governmental policy and programmes) and their nexus with environmental change form the backbone influencing migration and livelihood in the Sundarban Delta.

Perhaps, little is known how climate change could impact on gender relation through migration. Migration has been considered an element of women's agency if they have the opportunity to move independently; however, it is not sure if climate change could open up such possibilities. On the other hand, available studies throw more light on the women left behind. In the chapter on 'Gender processes in rural outmigration and socio-economic development in the Himalaya', Tiwari and Joshi observed that due to constraints of subsistence economy, a large proportion of male population outmigrates in search of livelihood from the Himalayas. Women are left behind looking after the mountain economy. The depletion of natural resources and climate change have further accelerated the trends of male outmigration, which have enhanced women's roles and responsibilities and increased their workload rendering them more vulnerable to environmental changes. Traditionally, women have a highly restricted ownership of natural resources and limited access to the opportunities of social and economic development. However, the situation started improving with an increasing role of remittances in the mountain economy. It also improved women's access to education, leadership and decision-making power. These changes are contributing towards social, economic and political empowerment of rural women. Furthermore, women have developed critical traditional knowledge to understand, visualize and respond to environmental changes including the climate change. Hence, it is highly imperative to improve rural livelihood in rural areas and extend the good practices of women mainstreaming in other areas across the Himalayan Mountains.

It is important to distinguish the impact of floods with that of the droughts. Floods often are sudden and catastrophic and the loss of income and lives is huge. Local people are displaced, immobilized and cattle die. In this situation, migrant family members provide great help to the families left behind affected by floods. Unlike floods, in a situation of drought, people are not displaced and immobilized but have the option to migrate. It is also possible for the government to initiate employment generation programme to curb migration from the drought affected areas. However, as the impact of drought is gradual, it has high potential to trigger migration (Findley 1994; Juelich 2011). Thus, drought deserves special mention in the study on climate change and migration relationship. Some parts of India experience frequent and severe droughts. In the chapter 'Climate change, drought and vulnerability: a historical narrative approach to migration from Western Odisha, India', Panda showed an increasing impact of climate change in recent years leading to heightened focus on climate-induced migration from different drought-prone areas of India. Although migration for livelihood diversification has existed as an adaptation strategy for farmers, increasing frequency of climate extremes is changing the contribution of climate as a sole variable for inducing migration. This chapter, based on household surveys and group discussions, tried to examine the monocausality between climate variability and migration from historical narrative perspective to tease out the role of climate in the overall setting of socio-economic factors in a drought-prone region in Eastern India. Historical narratives show how eventual eroding of financial and physical assets due to regular droughts played an important role in inducing migration on a regular basis. Further, the chapter examined the vulnerability of households and their links to migration and climate. The chapter argued that building assets of poor people who are at the threshold of slipping into chronic poverty is important in engaging households in remunerative migration as an adaptation to climate change rather than distressed migration. In another study, 'Dynamics of distress seasonal migration: a study of a drought-prone Mahabubnagar district in Telangana', Korra observed that the district of Mahabubnagar has been experiencing droughts frequently leading to an exodus of labour force from the district especially during the lean agricultural season. Seasonal migration is altering eventually, hence this study aimed at capturing the dynamics of distress seasonal migration. The study reveals that in the surveyed villages, people are increasingly becoming unemployable due to distress conditions and therefore witnessed surplus workforce. Squeezed economic opportunities, nonviable agriculture, high indebtedness, survival and people's increased monetary aspirations seems to be not just pervasive but mounting over the years. On the other hand, seasonal migration has failed to contribute to migrants' overall economic well-being, but brought solace to the migrant families during times of distress. In another chapter, 'Seasonal migration from dry climatic zone: a case of rural Maharashtra', Jaleel and Chattopadhyay also portrayed the similar story of distress seasonal migration from various parts of drought-prone areas of Maharashtra. In the district of Beed in Maharashtra, every year when dry season starts, thousands of peasant families undertake migration for working in sugar factories, brick kilns, stone quarries and much more, where they remain for varying periods generally until the rain starts again in their home areas. This chapter offers a glimpse into the life and work of seasonal migrant households, with an emphasis on the life and work of the seasonal migrant women. In a more precise way, the chapter looked into the characteristics, patterns, causes and consequences of seasonal migration of the rural households. Why these people migrate, how much of the year they are away from home, what types of activities they undertake at the destination, how they are paid and what they feel about their life are the major research questions addressed in this chapter.

A large-scale migration due to displacement is the direct result of the destruction of human ecology. In the chapter 'Migrant ecology', Banu describes the process of migration as embodied in the concept called 'migrant ecology'. It is fluidly conceived as the original habitat in which rural people live and search for their livelihood and also the new urban ecology into which they are forced to shift. When their rural ecology is destroyed due to developmental reasons, they migrate in search of new avenues. In the urban system, they are the most marginalized and live a life completely alien to them. Forced migrants now realize that a whole chain of hidden structure of intervention from multiple economic domains curbs their nurtured relation with their soil. Set under the condition of new developmental interventions, it redefines their sense of belonging to the ecosystem. The power of the authorities, who are part of the chain of the capitalist mode of production, needs to be understood not in isolation but as an overarching system which has a universal impact on our global ecosystem. The voices of the unknown poor migrant, 'other' from the below, then have to become the voices of all. Government's policy and programmes could play a very important role in protecting the rights of the migrants who are vulnerable to not only socio-economic and political forces, but also the natural forces unleashed by climate change. In the chapter 'Spaces of recognition of climate migrants in India: question of rights and responsibilities', Dey reminds the responsibility of the state and rights of migrants as citizens.

The author argues that climate change–related migration could evolve into a global crisis by displacing a large number of people from their homes and forcing them to flee. Based on Lefebvre's theory of space, this chapter highlights that climate migrants in India are creating their own space as they are pushed out and forced to survive. This chapter also examines the state of citizenship for the migrants as climate change–related migration in the age of globalization snatches the notion of 'national citizen' through spatial marginalization and socio-economic deprivations. In order to recognize the new citizenship in space, a new development model needs to be discussed and adopted which could promote social justice and protect the rights of the migrants.

Conclusion

This volume attempts to highlight the nature and pattern of migration in different parts of India affected by extreme climatic events like droughts, floods and cyclones potentially linked with climate change. Although droughts, floods and cyclones are not new in India's climatic history, their frequencies and intensities are new dimensions emphasized in the context of climate change. Assuming the wisdom of climate scientists, as summarized in the assessment reports of the IPCC, we believe that climate change is a reality, although the magnitude of change may vary in different parts of the world or even in different parts of a large country like India. In this context, different papers included in this volume show that climate change accentuates the existing socio-economic vulnerabilities. It is an important driver of migration but interwoven with other socio-economic and political factors. This volume indicates that climate change is a distant determinant of migration influencing through the proximate determinants of loss of income, employment opportunities and livelihood. As climate change is conceptualized as a push factor, the various chapters of this volume inform us about the dynamics of migration related to the areas of origin. Migration would occur as sudden like displacement or gradual depends upon the nature and intensity of extreme weather events. Further, remittances sent back to the household members played an important role in promoting adaption and resilience during the time of disaster and crisis. From a policy and programme point of view, there is a need to integrate migration and utilization of remittances with development and planning in both areas of origin and destination. This requires a change of mind-set of looking at migration as a part of solution instead of a part of the problem.

References

Adamo, S. B. (2009) *Environmentally Induced Population Displacements.* Bonn: IHDP Update I, International Human Dimensions programme on Global Environmental Change.

Bardsley, D. K. and G. J. Hugo (2010) "Migration and climate change: Examining thresholds of change to guide effective adaptation decision-making", *Population and Environment*, Vol. 32, No. 2–3, pp. 238–262.

Bates, D. C. (2002) "Environmental refugees? Classifying human migrations caused by environmental change", *Population and Environment*, Vol. 23, No. 5, pp. 465–477.

Bhagat, R. B. (2014) *Urban Migration Trends, Challenges and Opportunities in India*, Background Paper to World Migration Report 2015, IOM, Geneva.

Black, R. (1998) *Refugees, Environment and Development*. London: Longman.

Davis, J. and D. Lopez-Carr (2010) "The effects of migrant remittances on population-environment dynamics in migrant origin areas: International migration, fertility, and consumption in highland Guatemala", *Population and Environment*, Vol. 32, No. 2–3, pp. 216–237.

El-Hinnawi, E. (1985) *Environmental Refugees*. Nairobi: United Nations Environmental Programme.

Findley, S. E. (1994) "Does drought increase migration: A study of migration from rural Mali during the 1983–1985 drought", *International Migration Review*, Vol. 28, No. 3, pp. 539–553.

Foresight (2011) *Migration and Global Environmental Change*, Final Report, Government Office for Science, London.

Gomez, O. (2013) *Climate Change and Migration: A Review of the Literature*, a study commissioned by the International Institute of Social Studies. The Hague: Erasmus University.

Gray, C. and R. Bilsborrow (2013) "Environmental influences on human migration in rural Ecuador", *Demography*, Vol. 50, No. 4, pp. 1217–1241.

Gray, C. and V. Mueller (2012) "Drought and population mobility in rural Ethiopia", *World Development*, Vol. 40, No. 1, pp. 134–145.

Heberle, R. (1938) "The causes of rural-urban migration: A survey of German theories", *American Journal of Sociology*, Vol. 43, No. 6, pp. 932–950.

INCCA (2010) *Climate Change and India: A 4 X 4 Assessment: A Sectoral and Regional Analysis for 2030s*, INCCA Report No. 2, Ministry of Environment and Forests, Government of India, New Delhi.

International Organisation for Migration (IOM) (2015) *World Migration Report 2015, Migrants and Cities: A New Partnership to Manage Mobility*. Geneva: IOM.

IPCC (2012) *Managing the Risks of Extreme Events and Disasters to Advance Climate Change Adaptation (SREX)*. Special Report of the Intergovernmental Panel on Climate Change (IPCC). IPCC Secretariat, Geneva.

Juelich, S. (2011) "Drought triggered temporary migration in an east Indian village", *International Migration*, Vol. 49, pp. e189–e199.

Laczko, F. and C. Aghazaram (2009) *Migration, Environment and Climate Change: Assessing the Evidence*. Geneva: International Organisation for Migration (IOM).

Massey, D. S., W. G. Axinn and D. J. Ghimire (2010) "Environmental change and out-migration: Evidence from Nepal", *Population and Environment*, Vol. 32, No. 2–3, pp. 109–136.

Mazumdar, I., N. Neetha and I. Agnihotri (2013) "Migration and gender in India", *Economic and Political Weekly*, Vol. 48, No. 10, pp. 54–64.

Myers, N. (2002) "Environmental refugees: A growing phenomenon of the 21st century", *Philosophical Transactions (Royal Society)*, Vol. 357, No. 1420, pp. 609–613.

Piguet, E., A. Pécoud and P. De Guchteneire (eds) (2011) *Migration and Climate Change*. Paris, Cambridge: Cambridge University Press – Editions de l'UNESCO.

Raleigh, C., L. Jordon and I. Salehyan (2008) *Assessing the Impact of Climate Change on Migration and Conflict*. Washington, DC: The World Bank.

Rangarajan, C.and S. M. Dev (2014) "Counting the Poor: Measurement and Other Issues", Working Paper, WP-2014–04, Indira Gandhi Institute of Development Research, Mumbai December 2014 www.igidr.ac.in/pdf/publication/WP-2014-048.pdf.

Samson, J., D. Berteaux, B. J. McGill and M. M. Humphries (2011) "Geographic disparities and moral hazards in the predicted impacts of climate change on human populations", *Global Ecology and Biogeography*, Vol. 20, pp. 532–544.

Smit, B. and J. Wandel (2006) "Adaptation, adaptive capacity and vulnerability", *Global Environmental Change*, Vol. 16, pp. 228–292.

Stern, N. (2007) *The Economics of Climate Change: The Stern Review*. Cambridge: Cambridge University Press.

Trenberth, K. E., P. D. Jones, P. Ambenje, R. Bojariu, D. Easerling, A. Klein Tank, D. Parker, F. Rahimzadeh, J. A. Renwick, M. Rusticucci, B. Soden and P. Zhai (2007) "Observations: Surface and Atmospheric Climate Change", in Solomon, S., Qin, D., Manning, M., Chen, Z., Marquis, M., Averyt, K. B., Tignor, M. and Miller, H. (eds), *Climate Change 2007: The Physical Science Basis, Contribution of Working Group I to the Fourth Assessment Report of the Intergovernmental Panel on Climate Change*, L. Cambridge: Cambridge University Press, pp. 235–336.

UNDP (2009) *Human Development Report, Overcoming Barriers*. New York: Human Mobility and Development, United Nations Development Programme.

United Nations University (2007) "Environmental Refugees: The Forgotten Migrant", Panel Discussion, UN Head Quarters, New York, May 16, 2007 https://proyectoambientales.files.wordpress.com/2011/05/bogardi-2007-environmental-refugees_the-forgotten-migrants.pdf accessed on 16 December 2016.

Warner, K. (2011) "Environmental change and migration: Methodological considerations from ground-breaking global survey", *Population and Environment*, Vol. 33, No. 1, pp. 3–27.

2 Climate change, vulnerability and migration in India

Overlapping hotspots

R. B. Bhagat

India is one of the most vulnerable countries to climate change (Brenkert and Malone 2005). The Government of India is serious on this issue, as evident from the statement of the Ministry of Environment and Forests that 'no country in the world is as vulnerable, on so many dimensions, to climate change as India. Whether it is our long coastline of 7,000 km, our Himalayas with their vast glaciers, our almost 70 million hectares of forests (which incidentally house almost all of our key mineral reserves) – we are exposed to climate change on multiple fronts. Rigorous science based assessments are therefore critical in designing our adaptation strategies' (INCCA 2010: 9). In the 12th Five Year Plan (2012–2017), the Government of India has also proposed a National Mission on Strategic Knowledge for Climate Change which identified challenges arising from climate change, promotes the development and diffusion of knowledge on responses to these challenges in the areas of health, demography, migration and livelihood of coastal communities (Planning Commission 2011: 76).

The climate change is not a speculative concern but based on the works of hundreds of scientists under the aegis of IPCC. The UNEP and the WMO established IPCC which was endorsed by the United Nations General Assembly in 1988. Climate change is recognized by IPCC as a significant man-made global environmental challenge. International efforts to address climate change began with the adoption of the United Nations Framework Convention on Climate Change in 1992, following publication of first assessment report by IPCC in 1990. Until now, the IPCC has published four rounds of assessment reports and the fifth will be submitted in 2014.

It is to be noted that the science of climate change is still evolving. There are several aberrations which climate scientists are struggling to explain and there are many parameters on which there is no consensus. However, the point before climate science today is not whether man-made global warming is real, but whether the models being used to predict climate change are reliable enough to inform policymakers' decisions (Roulstone and Norbury 2013: 11). In addition to climate change, human activities also alter the uses of land, water and ecosystem services in such a way that benefits some and costs others. Both long- and short-term environmental changes bear enormous socio-economic consequences for livelihood security and well-being of the people (Foresight 2011: 26).

Climate change affects migration through a number of socio-economic and political drivers. Some of the people in the areas of climate change risk decide to migrate and also have the ability to migrate, while others are forced to stay back. The Foresight Group argues that this trapped population, that is, those who do not move, are equally important in adaption and planning for climate change–affected people (Foresight 2011). As climate change–induced migration is confounded by a number of socio-economic determinants, it is however difficult to predict. But the fact remains that a large number of people are getting affected by changes in the rising temperature and precipitation and also extreme weather events like floods, droughts, cyclone and cloudbursts. Desertification and declining crop productivity are slow processes affecting the livelihood of the people and forcing them to seek the option of migration. This may also raise mortality and morbidity levels in the affected population if mitigation and adaption strategies are not followed and may derail the efforts to achieve millennium development goals. Migration could be a possible adaptive mechanism, but as several studies on India show, migration is socio-economically a selective process (Oberai and Singh 1983; Skeldon 1986; Bhagat 2010). This means that not everybody affected by climate change would be privileged to migrate and settle elsewhere. However, this may likely increase the seasonal and temporary migration, may intensify impoverishment and malnourishment and increase the morbidity and mortality levels of vulnerable populations. There could be various possibilities, and migration is one that depends upon the number of socio-economic and political drivers. This paper identifies the areas of climate change vulnerabilities and the pattern of migration in India based on published sources and available data. Climate change vulnerability and migration are two uncertain outcomes that depend upon many factors such as the intensity of climate change, adaptation and opportunity of migration at various destinations. This paper also argues that climate change should not be seen merely as a biophysical change, but must be looked upon embedded in the socio-economic conditions of population affected by it. It means that populations differ in their adaptive capacity, and those who are socio-economically poor occupying the hotspots of climate change are likely to be severely affected. This paper highlights the spatiality of the correspondence between the areas of climate change vulnerability, socio-economic deprivations and the emerging pattern of migration in India.

Climate change and India's biophysical entity

Global climatic models fail to simulate fine regional features of India which is not only a vast country, but also having a huge climatic diversity. The INCCA's study shows that mean temperature in India increased by about 0.2 degree Centigrade per decade (i.e. 10 years) for the period 1971–2007, with a much steeper increase in minimum temperature than maximum temperature (INCCA 2010: 28). Other studies on climate change in India also point to clear increase in temperature over India, especially in winter with enhanced warming during night than day (Kumar et al. 2011) and over northern India (Kulkarni 2012). Also, this has an impact on the monsoon rainfall pattern which occurs in most parts of India from June to September. India is heavily dependent on agriculture, a source of livelihood of half the population. Seasonal mean rainfall shows a noticeably declining trend with more frequent deficit monsoons leading to decreased precipitation over central India along the monsoon trough (Kulkarni 2012). This is happening due to a number of factors such as increase in black carbon, sulphate aerosols (Chung and Ramanathan 2006; Bollasina et al. 2011) and land-use changes (Niyogi et al. 2010). Scientists have also observed an increase in the number of monsoon break days over India consistent with the overall decrease in seasonal mean rainfall (Dash et al. 2009). Also, the frequency of heavy precipitation events is increasing (Rajeevan et al. 2008; Krishnamurthy et al. 2009; Pattanaik and Rajeevan 2010), while light rain events are decreasing (Goswami et al. 2006). In a nutshell, monsoon has been behaving more erratically which is the life line of Indian agriculture as more than half of India's population, that is, 1.21 billion as per the 2011 Census, is dependent on agricultural activities.

Sea level rise might induce huge displacement and relocation of population from the coast to higher elevation. The mean sea level rise along the Indian coasts is estimated to be about 1.3 mm per year on an average (INCCA 2010: 47). Population in low elevation coastal areas is especially vulnerable to risks resulting from climate change as these lowlands are densely settled and growing rapidly. The rising sea level will increase the risk of floods, and stronger tropical storms may further increase the flood risk. Low-income groups living on flood plains are especially vulnerable. The eastern coast is more vulnerable, as a significant number of cyclones have been observed to have occurred in the last 100 years in the Bay of Bengal as compared to the Arabian Sea. According to the INCCA study, there are three coastal regions highly vulnerable, namely Nagapattinam in Tamil Nadu, which is characterized by a flat onshore topography; Paradeep in Orissa, an area with frequent occurrence of storm surges; and Kochi, a low-lying region characterized by the presence of backwaters (INCCA 2010: 16). The INCCA study also shows that fish production in the coastal areas which contributes to 5 per cent of India's agricultural GDP is being affected (INCCA 2010: 75). The region has experienced increased flooding recently due to increased deforestation along the Western Ghats (hills), soil erosion and changes in the intensity of monsoon rains (Working Group on Climate Change and Development 2007). There has also been a huge diminution of mangrove forest for which the coastal areas are known for. The largest mangrove forest, however, still exists in the Gangetic delta of the Sundarban in the east coast. In between the coastal areas and Deccan plateau, there exists a narrow strip of Western Ghats (hills) which is rich in biodiversity. It runs through several states of western and south India, spread over 160,000 square kilometres. About one-fourth of this region is under forests. The growth of population around protected areas and forests has led to increasing human-wildlife conflict (INCCA 2010: 19). The region has experienced an increased urbanization in recent years and also the rapid growth of tourism industry in this region. Scientists predict that this region is likely to experience rise in temperature, changes in the amount of precipitation and also number of rainy days with significant impact on the biodiversity (INCCA 2010: 129–143).

India has a long mountainous system known as Himalayas spread over from the state of Jammu and Kashmir in the west to Arunachal Pradesh in the east. It covers 16 per cent of country's geographical area and about 4 per cent of India's population. This means that about 50 million people who largely practise hill agriculture and whose vulnerability is expected to increase on account of climate change, practise hill agriculture and whose vulnerability is expected to increase on account of climate change.

The rising temperature is likely to affect the hydrological cycle in the Himalayas leading to melting of glaciers and increase in the frequency of floods and droughts (Foresight 2011: 92). The region has experienced huge land degradation and deforestation due to development activities in recent years. Although the region has been the attraction of tourists, much of the Himalayan area is characterized by a very low economic growth rate combined with a high rate of population growth (INCCA 2010: 16–17). As a result, with the region having limited mineral resources and modern industries, most of the population is dependent on agriculture and horticulture. Some of the areas of Himalayas such as the state of Uttarakhand have experienced persistent outmigration in the past.

Box 2.1   Government of India initiative on climate change response

Ministry of Environment and Forests, Government of India formally launched the Indian Network of Climate Change Assessment (INCCA) 2009. INCCA is a network-based programme that brings together over 120 institutions and over 220 scientists from across the country to undertake scientific assessments of different aspects of climate change. It emphasizes the need of developing the indigenous ability to measure, model and monitor (3 Ms) climate change, the foundation of India's decision-making on climate change.

Source: INCCA 2010

Unlike the coastal and mountainous region, the drylands or desert margins are extremely water scarce areas. About 40 per cent of earth's land surface is characterized as drylands which sustain over 2 billion people. While the direction of the climate change in the drylands is uncertain, most of the areas are exposed to the effects of land degradation and climate change (Foresight 2011: 68). In India, the dryland regions are spread in the north-west India predominantly confined to western Rajasthan. Some studies indicate the declining trend in precipitation in north-west India may likely increase the drought frequency in future (Trenberth et al. 2007: 255–256).

North-eastern India is another hotspot of biodiversity comprising seven sister states, namely Arunachal Pradesh, Assam, Manipur, Meghalaya, Mizoram, Nagaland and Tripura along with Sikkim. Ethnically, north-east India is linguistically and culturally very distinct from the other states of India. This region is officially recognized as a special category of states which constitute 8 per cent of the country's total geographical area and about 4 per cent of the total population of

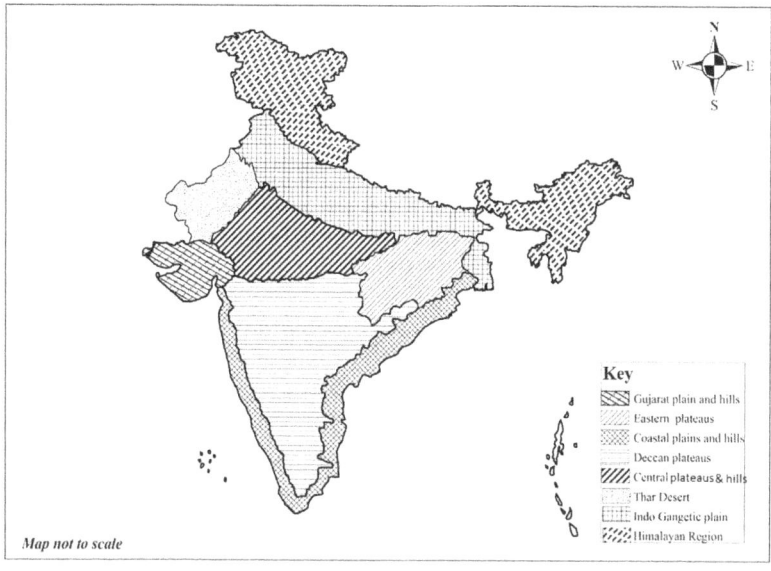

Map 2.1 Biophysical divisions of India

Source: Compiled by the author; see also Khanna (1989) Agro-climatic regional planning: an overview (unpublished). Planning Commission, New Delhi, p. 144

the country. The region has geopolitical significance as it also shares international borders with neighbouring countries. The states like Nagaland, Mizoram and Meghalaya have predominantly scheduled tribe population, and the whole region is dependent on agriculture and natural resources with 52 per cent of area under forests. The surface temperature is likely to increase by about 2°C by 2030 from the level of 1970 and region is likely to experience extreme changes in precipitation in the future (INCCA 2010: 41, 124).

The biophysical map of India is very complex and the above brief description of the macro biophysical regions intends to provide a background to understand the climate change vulnerability and emerging pattern of migration in India (see Map 2.1).

Theoretical framework

The neoclassical theories of migration (Lewis 1954; Harris and Todaro 1970; Todaro 1976), the new economics of migration (Stark and Bloom1985) and also sociological theories based on network and social capital could not foresee the effect of climate change on migration (Massey et al. 1993; de Hass 2008). However, during the last one decade, there have been two important changes in our thinking and knowledge of both climate change and migration. There is now greater awareness and concern on climate change among nations, global leaders and policymakers, and also a paradigm shift from pessimism to optimism about the role of migration in development (de Hass 2008; UNDP 2009). It is not true that migration does not have negative consequences, but there are equally and even larger positive consequences in some contexts. In addition to the role of migration in removing imperfections in the labour market, remittances both monetary and non-monetary like knowledge, attitude and information also play an important role in improving economic conditions and well-being of the people in the areas of origin (Deshingkar and Matteo 2012). However, it is possible to minimize the negative consequences of migration through planning and programmes if migration is viewed positively and integrated into the development programmes. In this context, it is not only important to know who migrates, but also who stays back. The Foresight Group (2011) identified stayers – a very important group in understanding the relationship between migration and climate change. There are two groups of stayers – those who want to migrate but lack the resources to migrate and those who do not wish to migrate. The former is called the trapped population by the Foresight Group. The Foresight Group also argued that preventing migration could be counterproductive as it increases the risk of climate change on the trapped population in the future. There is another group of migrants who do not want to migrate but are forced to migrate or displaced as a result of the occurrence of extreme weather events like floods, droughts, cloudbursts, cyclones and so on. Forced migration and displacement can also occur due to earthquakes and tsunamis as well, but perhaps may not fall under the purview of climate change.

As mentioned earlier, climate change influences migration through a number of socio-economic and political factors. The impact of climate change on human mobility could be either rapid as a result of extreme weather events like cyclones, storms, floods and so on or slow due to desertification, changes in the rainfall pattern, sea level rise and melting of the mountain glaciers. It is agreed in a number of studies that due to the interwoven and complex nature of the relationship between climate change and socio-economic and political factors, its influence on migration is difficult to be separated (Foresight 2011; Gomez 2013). Also, migration is one of several mobility outcomes along with displacement and movement of refugees. As mobility outcomes are varied and diverse, each outcome is influenced by the nature and pattern of climate change and socio-economic and political interactions differently. However, the key to understanding mobility outcomes lies in the opportunity or loss of livelihood and decision to migrate as influenced by vulnerability, aspiration, network and ability to migrate. An understanding of the complex and varied relationship between climate change and migration will be helpful in planning the movement and realizing the potential of migration as an adaptive means to climate change (see Fig. 2.1). Vulnerability is the key to the understanding of climate change and migration. Vulnerability is the likelihood that an individual or group will be exposed to the adverse impact of climate change and their capacity to reproduce themselves. Vulnerability could be divided into two categories: biophysical vulnerability and socio-economic vulnerability. While biophysical vulnerability is the result of climate change and the resulting extreme weather events, socio-economic vulnerability is the outcome of income level, poverty, educational level, social capital

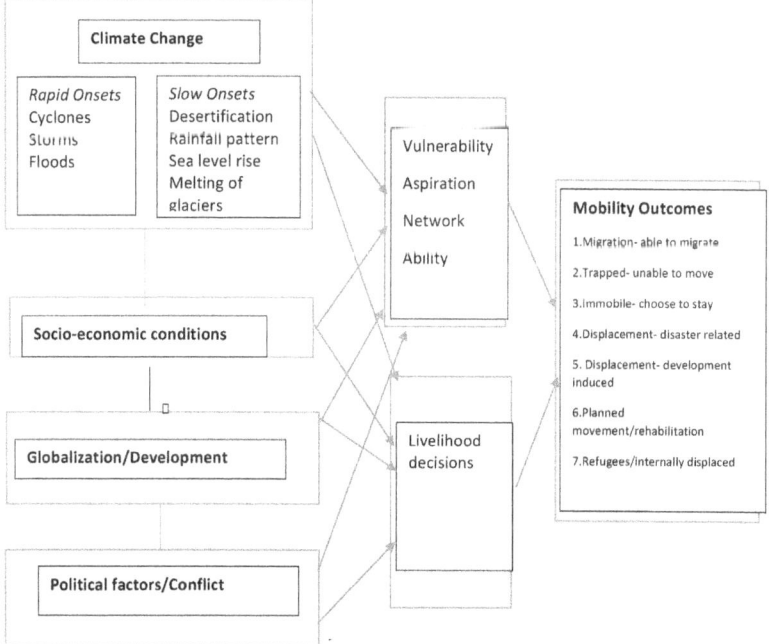

Figure 2.1 Conceptual framework depicting the relationship between climate change and mobility outcomes

Source: Created by the author

and network. The climate change vulnerability is the net outcome of both – the former related with the risk and the latter with the adaptive capacity. Migration predominantly influenced through socio-economic conditions is one of adaptive means among several alternatives to the sources of livelihood. While vulnerability is associated with risk, adaptation is the ability of the community to survive and reproduce themselves in a changing environmental context.

In the context of India, the climate change–induced biophysical vulnerability is likely to affect almost all macro regions of India, namely the Himalayas, dry areas, the Western Ghats and coastal areas, but the impact on migration will be mediated through socio-economic vulnerability to the climate change and the ability to migrate. The next section presents the emerging pattern of migration in different biophysical zones in India and also the likely impact of climate change on future migration pattern and displacement.

Climate change vulnerability: hotspots

The identification of hotspots of climate change vulnerability depends upon existing knowledge on the subject, which is growing in recent years (Anthony-Smith 2009). According to the Asian Development Bank (2009), low-lying coastal areas, deltaic regions and semi-arid areas are the hotspots vulnerable to climate change. Hotspots are defined as specific areas or regions that may be at relatively high risk of adverse impact of one or more natural hazards as a result of climate change. The west coast, the Ganges-Brahmaputra Delta, the delta of Mahanadi, Krishna and Godavari in the east coast and arid areas of Rajasthan are the hotspots. On the other hand, INCCA (2012) considered only four regions namely, coastal areas, Western Ghats, north-east and the Himalayas as hotspots of climate change vulnerability based on biodiversity, and left the western dryland which is an important biophysical entity of India. The Foresight Group (2011) identified drylands, mountainous region and low elevation coastal zones as areas of hotspots of climate change vulnerability. It is worthwhile to note that the climate change models are still rather imperfect representations of reality, and differ considerably in identifying the zones of vulnerability and hotspots (Ericksen et al. 2011: 13).

In a very significant study carried out by O'Brien and colleagues (2004) on India argues that climate change vulnerability is not only the product of biophysical changes, but also a function of a range of socio-economic factors. IPCC also suggested that vulnerability may be characterized as a function of three components, namely adaptive capacity, sensitivity and exposure. Adaptive capacity describes the ability of a system to adjust which is a function of income, technology, education, information, skills, infrastructure, access to resources and stability and management capabilities. Sensitivity, on the other hand, refers to the degree to which a system will respond to a change in climate, either positively or negatively. Exposure relates to the degree of climate stress upon a particular unit of analysis. Using this framework, O'Brien and colleagues (2004) prepared a map of climate change vulnerability of Indian agriculture at district level. Map 2.2 shows the current vulnerability to future climate change across districts of India.

The climate sensitivity index used in O'Brien's study primarily captured the extreme weather events like droughts and floods that occurred during 1961–1990. As the study did not incorporate climate change impact on sea level rise, many urbanized and relatively more industrial states with long coastal lines did not fall under the category of high climate change vulnerability. However, another significant study by Brenkert and Malone (2005), which takes into account the sea level rise, found six states more vulnerable than India as a whole, attributable largely to sensitivity to sea storm surges. These states are Goa, West Bengal, Kerala, Tamil Nadu, Orissa and Gujarat. All six

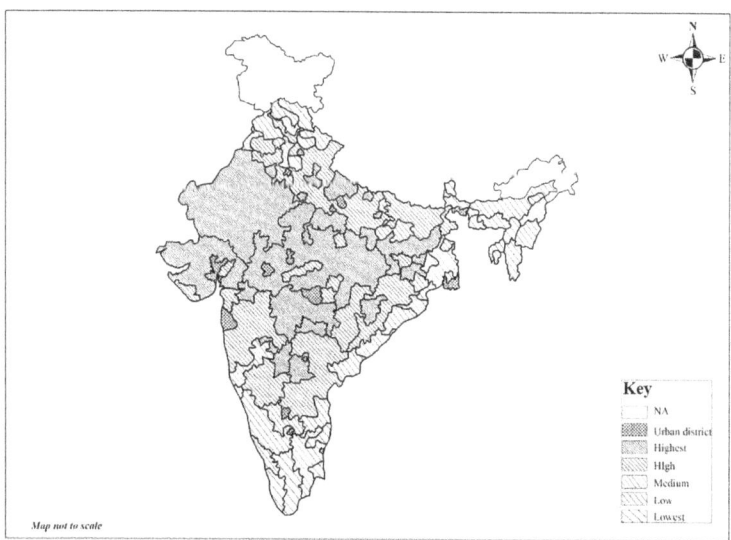

Map 2.2 District-level mapping of climate change vulnerability, measured as a composite of adaptive capacity and climate sensitivity under exposure to climate change (districts are ranked and presented as quintiles)

Source: Map adapted from O'Brien et al. (2004)

states are coastal states with high population density and higher level of urbanization except Orissa. In contrast, this study shows that mountain states exhibit highest resilience among Indian states, although they represent only low percentage (4%) of India's total population. Both studies highlight two important dimensions of India's vulnerability to climate change, that is, droughts and desertification, predominantly occurring in the arid and semi-arid areas of central and western India and the sea level rise affecting the coastal states.

Migration, displacement and climate change vulnerability: overlapping hotspots

Various rounds of decennial census in India until the 2001 Census show that about 30 per cent of India's population is migrants. However, as per the 2011 Census, migration has considerably increased from about 30 per cent in earlier censuses to 37 per cent in 2011. The total number of migrants enumerated was 453 million out of India's total population of 1,210 million in 2011. The majority of the migrants move within the states, whereas interstate migration constitutes only less than 15 per cent of the total migrants.

At state level, per capita income is strongly correlated with inmigration rate followed by percentage of workforce in non-agricultural sector and literacy rate. The pattern is consistent with the fact that developed states receive a large number of migrants from the relatively underdeveloped states (Bhagat 2010). The inland states like Punjab, Haryana and Delhi and many coastal states like Maharashtra, Gujarat, Goa, Karnataka, Kerala, Tamil Nadu and West Bengal have been receiving high level of inmigration in the past. These states are also more urbanized than the national average (31% urban in 2011) with Goa having the highest level of urbanization of 62 per cent followed by Tamil Nadu (48%), Kerala (47%), Maharashtra and Gujarat (45% and 42%), respectively. The urban population in India as a whole is growing at the rate of about 3 per cent per annum, whereas several coastal states are growing at a faster rate. On the other hand, many low urbanized states located in northern, central, eastern and north-eastern India are net outmigrating states. The emerging pattern of urbanization is shown in Map 2.3, which also indicates about the magnitude of rural-to-urban migration estimated to be as high as about 40 per cent of the urban growth in some of the most urbanized states (Bhagat and Mohanty 2009).

States could be broadly grouped keeping in view the biophysical consideration and socio-economic vulnerability, and the results are summarized in Figure 2.2. It may be noted that the mountainous areas,

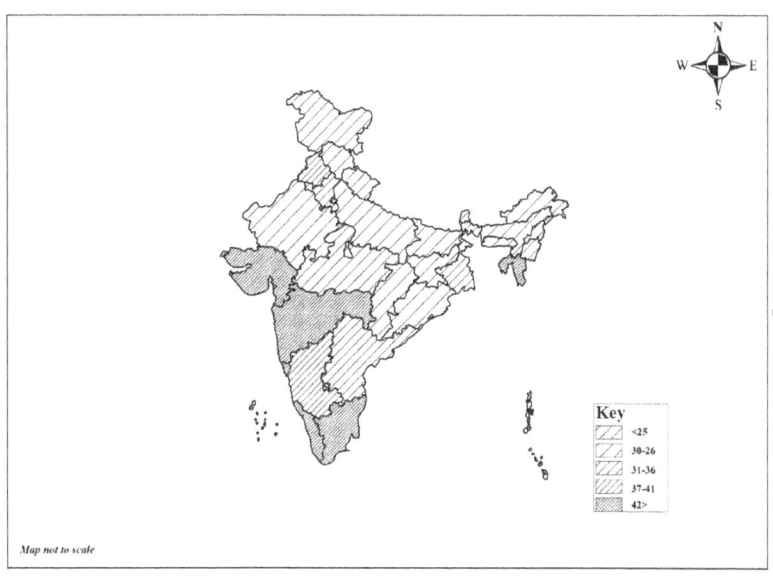

Map 2.3 Level of urbanization in India (in per cent), 2011

Source: Prepared by author

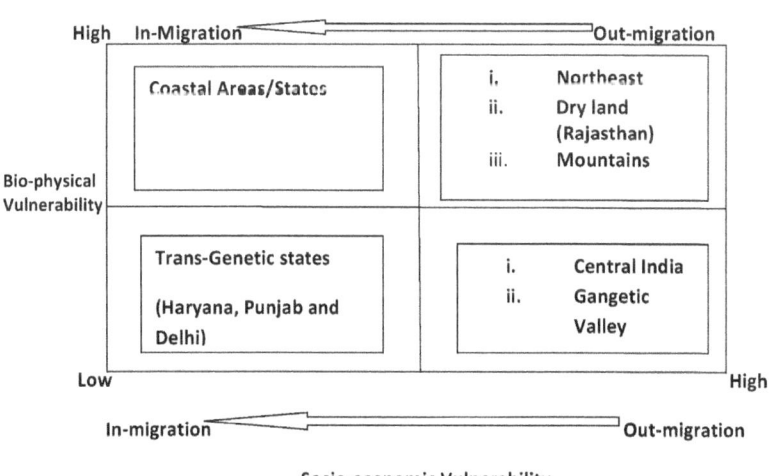

Figure 2.2 Regional pattern of climate change vulnerability and migration in India

Source: Created by the author

areas of the north-east, dryland of Rajasthan on the one hand and central India and most parts of the Gangetic valley on the other are sending migrants to the trans-Gangetic states of Punjab, Haryana and Delhi and the western and coastal states.

Migration takes place due to many reasons such as search for livelihood and employment, business, marriage, study, moved with earning member of the household, moved due to social and political conflicts and also as a result of natural disasters. Natural calamity like drought, flood, cyclones and more as a reason of migration was included in the 1991 Census, but was dropped in subsequent censuses perhaps because of its very low contribution. The reported rate was 5 per 1,000 migrants in 1991.

The NSSO also included natural disaster as a reason of migration. The 64th round pertaining to the year 2007–2008 gives an even lower figure, that is, 2 per 1,000 migrants.

If same rate is applied to the estimated migrants in 2011, the volume of migration due to natural calamity would be less than 2 million.

However, there are obvious interstate variations with Assam showing the highest rate of 51 per 1,000 migrants as a result of natural calamity. Also, in other north-eastern states like Arunachal Pradesh and Meghalaya, the prevalence was reported higher followed by the coastal states of Orissa (8), West Bengal (6) and Karnataka (6; see Fig. 2.3). In a statistical modelling of migration and related variables

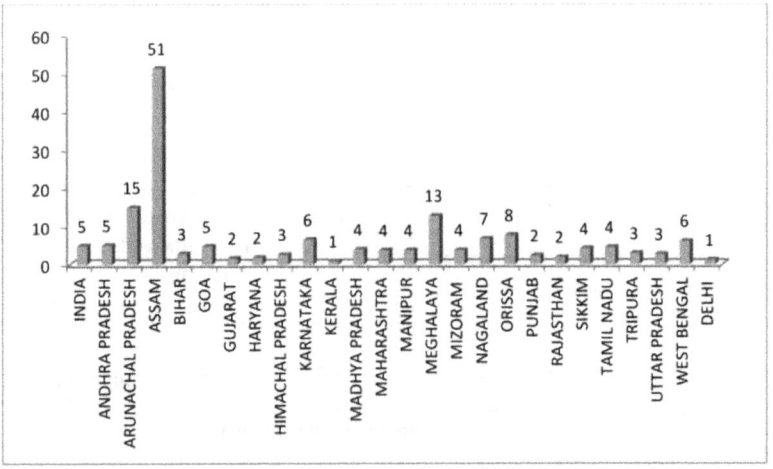

Figure 2.3 Natural calamity as a reason of migration per 1,000 migrants
Source: The 1991 Census of India

of droughts and floods migration is not sensitive to floods, whereas frequency of droughts has a low impact on interstate migration during the period 1991–2001 (Dallmann and Millock 2013). The low magnitude of migration due to natural calamity is due to the fact that it can capture only that part of migration which is caused by rapid onsets in climate changes.

Apart from rapid onsets, slow onsets like desertification, changes in the pattern of rainfall, sea level rise, receding of mountain glaciers affect various socio-economic drivers of migration, which are not captured by the direct question asked in the Censuses and National Sample Surveys.

When natural calamity as a reason of migration is compared with other reasons of migration based on census data, we find that the proportion of migrants moving as a result of work, employment and business constitute about 10 per cent of the total migrants (31% among males and less than 2% among females) as per the Census 2001. One of main criticisms of census data pertains to the fact that it is not able to capture seasonal and temporary migration.

There is also a large number of people circulating seasonally and temporarily for livelihood purposes, say for example, from one month up to six months in a year. The estimates of seasonal and temporary migration vary from about 15 million to 100 million in different studies (UNESCO and UNICEF 2012: 5). This type of migration is largely a livelihood strategy that results from the lack of adequate employment locally and is a means of coping with the household risks often associated with monsoon or crop failures (Keshri and Bhagat 2012). Longitudinal studies at village level also substantiate increased temporary migration due to rainfall shocks (Badiani and Safir 2008).

About half of India's 1.21 billion people depend on agriculture for their livelihood, although agriculture contributes only less than 15 per cent of the GDP. So, the income level of the majority of the households dependent on agriculture is extremely low. On the other hand, performance of Indian agriculture depends upon the behaviour of monsoon which lasts between June and September. Indian agriculture is likely to have considerable adverse impacts due to climate change leading to a significant drop in yields of important cereal crops like rice and wheat under various climate change scenarios, which could trigger migration of people dependent on agriculture for their livelihood (Kumar and Parikh 2001; World Bank 2008).

A large part of the country falls under semi-arid conditions with very low and erratic rainfall. Figure 2.3 shows that areas which are most vulnerable to climate change falls under central and western India consisting of arid, semi-arid and subhumid regions followed by upper and middle Gangetic valley.

Nomadic migration, although short distance, was an important feature outside the Gangetic valley (Hutton 1986: 61). This practice is still found in parts of Rajasthan and Madhya Pradesh located in central India. A number of studies also noted the short duration migration from the arid and semi-arid areas (O'Brien et al. 2004; Deshingkar 2005). The emerging map of climate change vulnerability of India's agriculture shows the states of Rajasthan, Gujarat, Madhya Pradesh, Chhattisgarh, Jharkhand, highland Odisha, several pockets in Uttar Pradesh and Bihar, Vidharbha and Marathwada in Maharashtra, Telangana and north-east Karnataka are extremely vulnerable to climate change. The available study on migration shows that a large part of this region has already been experiencing huge temporary and seasonal labour migration due to lack of livelihood opportunity (see Map 2.4). The average seasonal and temporary labour migration is 21 per 1,000 in the age group 15–64. However, the five areas, namely southern Rajasthan and south-eastern Gujarat; southern region of Uttar Pradesh, Vindhya and southern region of Madhya Pradesh; Bihar, northern part of Jharkhand and parts of West Bengal; south-eastern Odisha; and

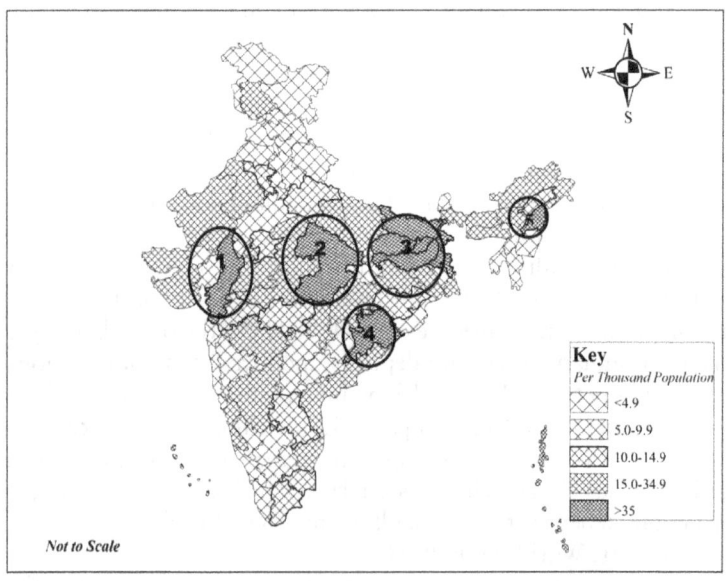

Map 2.4 Seasonal and temporary migration in India, 2007–2008

Source: Adapted from Kunal Keshri (2013)

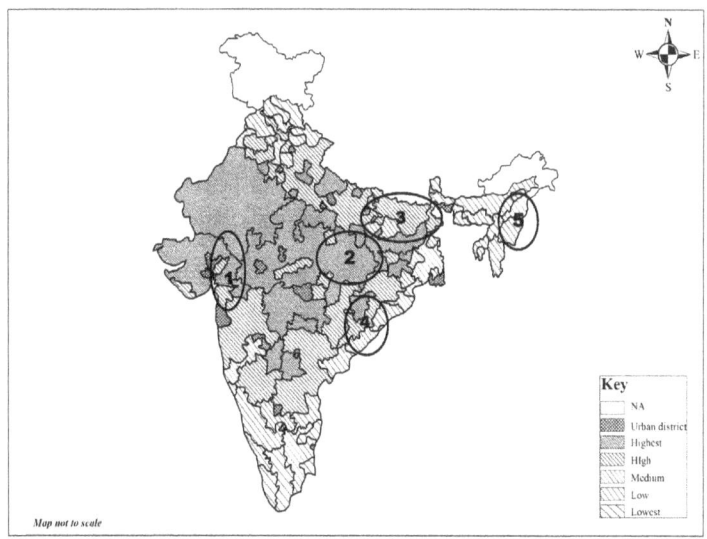

Map 2.5 Climate change vulnerability and hotspots of migration (1–5) – district-level mapping of climate change vulnerability of India's agriculture sector, measured as a composite of adaptive capacity and climate sensitivity under exposure to climate change

Source: Adapted from O'Brien et al. (2004) and Kunal Keshri (2013)

Note: Numbers 1–5 in circle are hotspots of temporary and seasonal migration

Nagaland in northeast India, where seasonal and temporary migration rate is estimated to be more than 35 per 1,000 population in the age group 15–64. The five regions showing relatively high seasonal and temporary migration could be termed as hotspots of migration which fall in the zone of high climate change vulnerability (see Fig. 2.8). Also, to note, seasonal and temporary labour migration is seven times larger than long-term labour migration in the age group 15–64, and a majority of them head towards urban areas (Keshri and Bhagat 2013).

Box 2.2 Hotspots of seasonal and temporary migration

1   The southern region of Rajasthan and south-eastern region of Gujarat; this region of Gujarat is the tribal belt with a history of seasonal migration (see also Breman 1994).

2   The Bundelkhand of Uttar Pradesh and Vindhya and south-
    ern region of Madhya Pradesh.
3   The largest pocket of higher prevalence of temporary migra-
    tion lies in eastern India consisting of northern and southern
    regions of Bihar, Hazaribagh Plateau of Jharkhand and the
    eastern plains of West Bengal.
4   The south-eastern part of southern Odisha.
5   Nagaland represents another centre of high prevalence of
    temporary labour.

The hotspots of high temporary labour migration are located
mostly in dry and hilly areas with a high proportion of tribal
population and consist of some of the most backward and poorest districts of the country.

Source: Kunal Keshri (2013)

Sea level rise increases a significant threat to human habitation in
the coastal areas. India has a long but densely populated coastal areas
spread over in the states of Gujarat, Maharashtra, Goa, Karnataka, Kerala, Tamil Nadu, Andhra Pradesh, Orissa and West Bengal. The Union
Territory of Daman and Diu, Dadra and Nagar Haveli and Pondicherry
also fall in this region. India's prominent mega cities like Mumbai, Kolkata and Chennai are located in this region. Also, there are a large
number of industrial, commercial and tourist centres like Surat, Mumbai, Panaji, Mangalore, Trivandrum, Kochi, Vishakhapatnam and Puri
located in the coastal zone. As the region is more urbanized with a
number of vibrant urban and port centres, it has also been receiving
a large number of migrants from both within and outside the region.
Flooding is common in many cities along the major rivers, which flow
from the mainland into the sea. The devastating flood of 26 July 2005
in Mumbai is still fresh in the memory of many people (Bhagat et al.
2006). The coastal habitation being affected by storm surges and the
greater salt water intrusion into the ground and surface water bodies is
likely to affect a large number of people due to high population density,
higher urbanization and migration in the region.

The low elevation coastal zone (i.e. area located between 0 m and
10 m above sea level) harbours about 3 per cent of land area and 6 per
cent of India's population (McGranahan et al. 2007), that is, about
73 million as per 2011. Myers (2001) argues that sea level rise could

displace 20 million people in India by 2050 (quoted in the Asian Development Bank 2009: 22). Greenpeace India estimates that 24 million people will be displaced if sea level rises to 1 metre and about 34 million if sea level rises to 3 metres by the end of the century (Greenpeace 2008: 10). Although the magnitude of displacement in coastal areas differs in various studies, the sheer size in millions is likely to throw serious challenges to urban centres. On the other hand, the impact of climate change on urban centres and migration as a response to climate is little recognized; as a result, urban planning is not sensitive to climate change and migration is not considered as a means of adaptation (Bhatt 2014).

Several studies recognize that it is not easy to predict migration and displacement because of their uncertain and interwoven nature with various socio-economic and political factors. However, they agree that migration would continue to happen and the environmental factors would keep on influencing the movements (Asian Development Bank 2009; Warner 2010; Foresight 2011; Centre for American Progress 2012; Gomez 2013). Also, the majority of climate change–induced migration will be within countries (Asian Development Bank 2009: 12), and most likely will exacerbate the existing migration pattern more than create entirely new flows (Barnett and Webber 2009: 17). However, it is important to mention that very few studies have focused on the relationship between climate change vulnerability and migration in India (Jha 2013; Dallmann and Millock 2013; Viswanathan and Kumar 2012; Greenpeace 2008; Badiani and Safir 2008). That not enough work is done points to the fact that both climate change and migration are uncertain outcomes and we need adequate and robust data on both aspects. It may be noted that there is a serious problem of lack of data on various forms of migration in India. Also, the available data provides information by large administrative units like states and districts. Further, migration data from the recently held 2011 Census are not yet available, and the latest available data from the NSSO pertains to the year 2007–2008.

Climate change impact on mobility should also take into account the immobility of population as well, that is, those who are most vulnerable but unable to migrate as migration is expensive and risky. When environmental change disrupts the livelihoods and ecosystem services, the most vulnerable are trapped while some better off may be able to move (Black et al. 2011). An understanding of this dimension is very important from policy and programme point of view, as the population left behind in the areas of origin vulnerable to climate change also need support as much as migrants in the place of destination.

The magnitude of the immobile population unable to move can be gauged by the magnitude of poor people who are not even able to meet their daily requirement of food. According to the recent figure released by Planning Commission (2013), India has 269 million people living below poverty line (21%), a majority of them found in the most vulnerable areas in the arid, semi-arid and subhumid areas of western, central and eastern India. As mentioned earlier, agriculture in this region is most vulnerable to climate change where people are predominantly dependent on agriculture for their livelihood. In the coastal states, about 80 million people live below poverty line, out of which half of them were found in six coastal states – Goa, West Bengal, Kerala, Tamil Nadu, Orissa and Gujarat – most vulnerable to the sea level rise. This group of people finds it difficult to migrate to places doing well, either the capital city or other prosperous areas (World Bank 2008: XIII).

Development policies will be better and inclusive if we integrate migrants in our development and urban planning and also take cognizance of the links between climate change and migration. Migration is beneficial to both areas of destination as well as areas of origin (UNDP 2009; World Bank 2009). Migrants' contribution to the skilled and unskilled workforce and their hard work is well documented. They are also engaged in the most difficult and risky jobs (3D jobs: dirty, dangerous and demeaning). On the other hand, remittances sent by them help families in improving their well-being in the place of origin. It is argued that preventing migration would be counterproductive (Tacoli 2009; Foresight Group 2011). The policies and programmes should integrate and support migrants as there is a widespread discrimination and exclusion of poor and unskilled migrants in social protection programmes due to lack of identity and residential proofs. Some are also discriminated in the schooling of their children, access to housing and work based on domicile (Bhagat 2011). As climate change migrants are one of the poorest, the existing discrimination might be reinforced if we do not recognize them in our development planning in general and urban planning in particular.

Conclusions

The biophysical structure of India is very complex. The mountains, hills, plateaus, plains, arid and semi-arid areas are interspersed with enormous variations in biodiversity. This makes an encompassing biophysical classification of India very difficult, at least at the macro and meso levels. As biophysical boundaries do not correspond with *tehsil*, district and state boundaries, socio-economic data are not easily available from the existing sources like the Census and the NSSO. This poses serious difficulties in the study of climate change and its consequences on migration. Also, data on various forms of migration are not up to date. In spite of difficulties and limitation of data, various climatic models have been applied on India to assess its regional vulnerability. Few studies have also attempted to examine the climate-related variables like changes in temperature and precipitation, occurrence of droughts and floods and their impact on migration through statistical models. This study summarizes the existing research works and attempts to identify the overlapping hotspot areas of climate change and migration in India. The study identifies two types of hotspots of climate change vulnerability. One that is mainly affected by its vulnerability to agriculture due to changes in the temperature and precipitation falls in the mainland of the arid, semi-arid and subhumid areas of western, central and eastern India. Others are the coastal states. Out of nine coastal states, six were identified most vulnerable to climate change due to rise in the sea level. The study shows that the areas which are most vulnerable to agriculture are also socially and economically poor, and a huge number of households practise seasonal and temporary migration for their livelihood. As a majority of the households are dependent on agriculture, the existing pattern of migration is likely to intensify in the future. The study identifies five such hotspots of migration located in the most vulnerable areas of climate change affecting India's agriculture. On the other hand, most of the vulnerable coastal states are more urbanized and are in migrating areas. These states are also likely to experience a huge displacement due to sea level rise up to 34 million by the end of the century. This study shows that the impact of climate change interacts with socio-economic conditions of the people in producing vulnerability. This is consistent with the Stern review that the low-income countries are more vulnerable to climate change than high-income countries (Stern 2007). As India has a huge number of poor people, this points to the fact that there is a larger number of trapped population who are unable to migrate compared to those who actually migrated. It must be emphasized that migration is an adaption to climate change, prevention could intensify vulnerability and even be more harmful. On the other hand, as most of the migrants move towards urban areas, urban planning sensitive to climate change and inclusive of migrants will be helpful in dealing with the emerging challenges.

Acknowledgements

I would like to thank Mr. Rakesh Kumar, PhD scholar and Mr Shushant Mandal, MPhil student at the International Institute for Population Sciences (IIPS), Mumbai, India, for helping me in cartographic work.

References

Anthony-Smith (2009) *Sea Level Rise and the Vulnerability of Coastal Peoples*, No. 7/2009. Bonn: UNU Institute for Environment and Human Security (UNU-EHS).

Asian Development Bank (2009) Addressing *Climate Change and Migration in Asia and Pacific: Executive Summary*. Manila: ADB.

Badiani, R. and A. Safir (2008) *Coping with Aggregate Shocks: Temporary Migration and Other Labor Responses to Climatic Shocks in Rural India.* New Haven, CT: Economic Growth Center, Yale, and Paris School of Economics (LEA).

Barnett, J. and M. Webber (2009) *Accommodating Migration to Promote Adaptation to Climate Change*, A Policy Brief prepared for the Secretariat of the Swedish Commission on Climate Change and Development and the World Bank Report 2010 Team.

Bhagat, R. B. (2010) "Internal migration in India: Are the underprivileged migrating more", Asia *Pacific Population Journal*, Vol. 25, No. 1, pp. 31–49.

Bhagat, R. B. (2011) "Migrants' (denied) right to the city", in *Urban Policies and the Right to the City in India: Rights, Responsibilities and Citizenship.* New Delhi: UNECSO and Centre de Sciences Humaines, pp. 48–57.

Bhagat, R. B., M. Guha and A. Chattopadhyay (2006) "Mumbai after 26/7 deluge: Issues and concerns in urban planning", *Population and Environment*, Vol. 27, pp. 337–349.

Bhagat, R. B. and S. Mohanty (2009) "Emerging pattern of urbanization and the contribution of migration in urban growth in India", *Asian Population Studies*, Vol. 5, No. 1, pp. 5–20.

Bhatt, S. (2014) (Available online at: www.indiatogether.org/sikkim-climate-change-book-reviews, accessed on 23/02/2014).

Black, R., W. N. Adger, N. W. Arnell, S. Dercon, A. Geddes and D. Thomas (2011) "The effect of environmental change on human migration", *Global Environmental Change*, Vol. 21, pp. S3–S11.

Bollasina, M. A., Y. Ming and V. Ramaswamy (2011) "Anthropogenic aerosols and the weakening of the south Asian summer monsoon", *Science*, Vol. 334, pp. 502–505.

Breman, J. (1994) *Wage Hunters and Gatherers: Search for Work in the Urban and Rural Economy of South Gujarat.* New Delhi: Oxford University Press.

Brenkert, A. L. and E. L. Malone (2005) "Modeling vulnerability and resilience to climate change: A case study of India and Indian states", *Climatic Change*, Vol. 72, pp. 57–102.

Centre for American Progress (2012) *Climate Change, Migration and Conflict in South Asia: Rising Tension and Policy Options across the Sub-Continent. Centre for American Progress*. Washington, DC: Centre for American Progress.

Chung, C. E. and V. Ramanathan (2006) "Weakening of North Indian SST gradients and the monsoon rainfall in India and the Sahel", *Journal of Climate*, Vol. 19, pp. 2036–2045.

Dallmann, I. and K. Millock (2013) *Climate Variability and Internal Migration: A Test on Indian Inter-State Migration*, University of Paris and Centre d'Economie de la Sorbonne.

Dash, S. K., M. A. Kulkarni, U. C. Mohanty and K. Prasad (2009) "Changes in the characteristics of rain events in India", *Journal of Geophysical Research-Atmospheres*, Vol. 114, No. D10, doi: 10.1029/2008JD010572.

deHass, H. (2008) *Migration and Development: A Theoretical Perspective*, Working Paper 9. Oxford: International Migration Institute, University of Oxford.

Deshingkar, P. (2005) "Maximizing the benefits of internal migration for development", in F. Laczko ed., *Migration and Poverty Reduction in Asia, International Organisation for Development*, Department for International Development, UK, and Ministry of Foreign Affairs, People's Republic of China.

Deshingkar, P. and S. Matteo (2012) "Migration and human development in India: New challenges and opportunities", in *National Workshop on Internal Migration and Human Development in India, Workshop Compendium Vol. II: Workshop Papers*. New Delhi: UNESCO and UNICEF, pp. 48–85.

Ericksen, P. P., A. Thornton, Notenbaert, L. Cramer, P. Jone, and M. Herrero (2011) *Mapping Hotspots of Climate Change and Food Insecurity in the Global Tropics*. CCAFS Report No. 5. Copenhagen, Denmark: CGIAR Research Program on Climate Change, Agriculture and Food Security (CCAFS). (Available online at: www.ccafs.cgiar.org).

Foresight (2011) *Migration and Global Environmental Change*, Final Report. London: Government Office for Science.

Greenpeace (2008) *Climate Migrants in South Asia: Estimates and Solutions*. Chennai: Greenpeace India Society.

Gomez, O. (2013) "Climate change and migration: A review of the literature", A study commissioned by the International Institute of Social Studies, The Hague (Erasmus University Rotterdam), within the project on 'Migration, Gender and Social Justice', funded by the International Development Research Centre (Canada).

Goswami, B. N., V. Venugopal, D. Sengupta, M. S. Madhusoodanan and P. K. Xavier (2006) "Increasing trend of extreme rain events over India in a warming environment", *Science*, Vol. 314, pp. 1442–1445.

Harris, J. and M. Todaro (1970) "Migration, unemployment and development: A two-sector analysis", *American Economic Review*, Vol. 40, pp. 126–142.

Hutton, J. H. (1986) *Census of India 1931 with Complete Survey of Tribal Life and System*, Vol. 1. Delhi: Gian Publishing House.

INCCA (2010) *Climate Change and India: A 4 X 4 Assessment: A Sectoral and Regional Analysis for 2030s*, INCCA Report No. 2. New Delhi: Ministry of Environment and Forests, Government of India.

Jha, A. K. (2013) "Climate change and internal migration in India: Response of the state, market and civil society", *Poverty & Public Policy*, Vol. 5, No. 2, pp. 133–145.

Keshri, K. (2013) "Regional pattern of temporary labour migration in India", *Geography and You*, (November–December 2013), pp. 54–58.

Keshri, K. and R. B. Bhagat (2012) "Temporary and seasonal migration: Regional pattern, characteristics and associated factors", *Economic and Political Weekly*, Vol. 47, No. 4 (January 28), pp. 74–81.

Keshri, K. and R. B. Bhagat (2013) "Socio-economic determinants of temporary labour migration in India: A regional analysis", *Asian Population Studies*, Vol. 9, No. 2, pp. 175–195.

Krishnamurthy, C. K. B., U. Lall and H. H. Kwon (2009) "Changing frequency and intensity of rainfall extremes over India from 1951 to 2003", *Journal of Climate*, Vol. 22, pp. 4737–4746.

Kulkarni, A. (2012) "Weakening of Indian summer monsoon rainfall in warming environment", *Theoretical and Applied Climatology*, doi: 10.1007/s00704-012-0591-4.

Kumar, K., S. Kavi and J. Parikh (2001) "Socio-economic impacts of climate change on Indian agriculture", *International Review of Environmental Strategies*, Vol. 2, No. 2, pp. 277–293.

Kumar, K., S. Patwardhan, A. Kulkarni, K. Kamala, K. Rao and R. Jones (2011) "Simulated projections for summer monsoon climate over India by a high-resolution regional climate model (PRECIS)", *Current Science*, Vol. 101, pp. 312–326.

Lewis, W. A. (1954) "Economic development with unlimited supplies of labor", *Manchester School of Economic and Social Studies*, Vol. 22, pp. 139–191.

Massey, D. S., J. Arango, G. Hugo, A. Kouaouci, A. Pellegrino and J. E. Taylor (1993) "Theories of international migration: A review and appraisal", *Population and Development Review*, Vol. 19, No. 3, pp. 431–466.

McGranahan, G., D. Balk and B. Anderson (2007) "The rising tide: Assessing the risk of climate change and human settlements in low elevation coastal zones", *Environment and Urbanisation*, Vol. 19, No. 1, pp. 17–37.

Myers, N. (2001) "Environmental Refugees – A Global Phenomenon of the 21st Century", *Philosophical Transactions of the Royal Society: Biological Sciences*, Vol. 357, pp. 167–182.

Niyogi, D., C. Kishtawal, S. Tripathi and R. S. Govindaraju (2010) "Observational evidence that agricultural intensification and land use change may be reducing the Indian summer monsoon rainfall", *Water Resources Research*, Vol. 46, No. 3, doi: 10.1029/2008WR007082.

Oberai, A. S. and H. K. M. Singh (1983) *Causes and Consequences of Internal Migration: A Study in the Indian Punjab*. New Delhi: Oxford University Press.

O'Brien, K., R. Leichenko, U. Kelkar, H. Venema, C. Aandahl, H. Tompkins, A. Javed, S. Bhadwal, S. Barg, L Nygaard and J. West. (2004) "Mapping vulnerability to multiple stressors: Climate change and globalization in India", *Global Environmental Change*, Vol. 14, pp. 303–313.

Pattanaik, D. R. and M. Rajeevan (2010) "Variability of extreme rainfall events over India during southwest monsoon season", *Meteorological Applications*, Vol. 17, pp. 88–104.

Planning Commission (2011) *Faster, Sustainable and More Inclusive Growth: An Approach to 12th Five Year Plan (Draft)*. New Delhi: Government of India.

Planning Commission (2013) *Press Note on Estimates of Poverty, 2011–12*. New Delhi: Government of India.

Rajeevan, M., J. Bhate and A. K. Jaswal (2008) "Analysis of variability and trends of extreme rainfall events over India using 104 years of gridded daily rainfall data", *Geophysical Research Letters*, Vol. 35, No. 18, doi: 10.1029/2008GL035143.

Roulstone, I. and J. Norbury (2013) "Climate-model misgivings", *Mint*, Vol. 7, No. 51, p. 11, (Available online at: www.livemint.com).

Skeldon, R. (1986) "On migration patterns in India during the 1970s", *Population and Development Review*, Vol. 12, No. 4, pp. 759–779.

Stark, O. and D. E. Bloom (1985) "The new economics of labor migration", *The American Economic Review*, Vol. 75, No. 2, pp. 173–178.

Stern, N. (2007) *The Economics of Climate Change: The Stern Review*. Cambridge: Cambridge University Press.

Tacoli, C. (2009) "Crisis or adaptation? Migration and climate change in a context of high mobility", *Environment & Urbanization*, Vol. 21, No. 2, pp. 513–525.

Todaro, M. P. (1976) *Internal Migration in Developing Countries*. Geneva: International Labor Office.

Trenberth, K. E., P. D. Jones, P. Ambenje, R. Bojariu, D. Easerling, A. K. Tank, D. Parker, F. Rahimzadeh, J. A. Renwick, M. Rusticucci, B. Soden and P. Zhai (2007) "Observations: Surface and atmospheric climate change", in S. Solomon, D. Qin, M. Manning, Z. Chen, M. Marquis, K. B. Averyt, M. Tignor and H. L. Miller, eds., *Climate Change 2007: The Physical Science Basis, Contribution of Working Group I to the Fourth Assessment Report of the Intergovernmental Panel on Climate Change*. Cambridge: Cambridge University Press, pp. 235–336.

UNDP (2009) *Human Development Report Overcoming Barriers: Human Mobility and Development*. New York: United Nations Development Programme.

UNESCO and UNICEF (2012) *National Workshop on Internal Migration and Human Development in India Workshop Compendium Vol. I: Workshop Report*. New Delhi: India Country Office.

Viswanathan, B. and K. K. S. Kumar (2012) "Weather variability, agriculture and rural migration: Evidence from state and district level migration in India",

Paper submitted for presentation at UNU-WIDER conference on Climate Change and Development Policy held in Helsinki on 28–29 September 2012.

Warner, K. (2010) "Global environmental change and migration: Governance challenges", *Global Environmental Change*, Vol. 20, pp. 402–413.

Working Group on Climate Change and Development (2007) *Up in Smoke: Asia and the Pacific*, Fifth Report. London: New Economics Foundation.

World Bank (2008) *Climate Change Impacts in Drought and Flood Affected Areas: Case Studies in India*, Report No. 43946-IN. New Delhi: Social, Environment and Water Resources Management Unit, India Country Management Unit, South Asia Region.

3 Migrating to adapt?

Exploring the climate change, migration and adaptation nexus

Himani Upadhyay and Divya Mohan

It has long been recognized that changes in the environment can influence human movement patterns and behaviour. Human migration in response to change in environment has been one of the considered strategies of the vulnerable households, to move away from the area of risk (McLeman and Smith, 2006). For nomadic and pastoralist communities, seasonal movement is an essential part of their livelihood. Migration induced by climate change was noted as early as 1990 by the IPCC (Intergovernmental Panel on Climate Change) which emphasized that 'the greatest single impact of climate change could be on human migration', with millions of people displaced by shoreline erosion, coastal flooding and severe drought. Since then, there have been various estimates which report that climate change will be one of the key drivers of population movement and displacement. But the focus has been on numbers. The estimates range from 200 million (Myers, 2005) to 1 billion (Christian Aid, 2007). The figure by Meyers has become the generally accepted figure, even though it has almost little empirical basis (Brown, 2008). Similarly, Lambert (2002) has estimated that there will be 20 million people displaced by climate change in China, though this is also supported by little empirical evidence. The Stern review noted that 'Greater resource scarcity, desertification, risks of droughts and floods, and rising sea levels could drive many millions of people to migrate' (Stern, 2006). When the International Organization for Migration (IOM) published that in 50 years there could be as many as 200 plus million environmental migrants (IOM, 2008; Warner, 2010), media, public and research interest in the subject multiplied instantly. The media interest with issue led to reports from all across the world forecasting widespread migration of vulnerable population fleeing their homeland (Bhagat, 2009; Sherriff, 2005; Bulman, 2005). These developments have led to debates and controversies regarding the climate change and migration topic (Hartmann, 2010). Most of the discussions revolve around 'how' many numbers of migrants, 'where' will it happen, and 'what' would be the consequences. 'Why' it happens, beyond climate change but exploring climate change in wider contexts, is explored less often.

The work done on climate change, migration and adaptation broadly falls under two categories: one that does not consider migration as adaptation but rather an outcome of failure to adapt; and second, which promotes migration as an important adaptation strategy. The estimates cited above reaffirm the first broad category, wherein the figures have certain underlying assumptions that presume that migration reflects the failure to adapt to changes in physical environment. This position reflects a negative connotation wherein migration is a forced option. Predominance of this point of view is also exemplified in the absence of mobility as an adaptation strategy in the cases collected under the UNFCCC database on distribution of different kinds and combinations of local coping strategies and adaptation practices (UNFCCC, 2014; Agrawal and Perrin, 2009). The second category of work has opposite and positive view on migration where it is a chief adaptive response to socio-economic, cultural and environmental change. It also highlights that migration, when planned and voluntary, can serve as an essential coping strategy for addressing climate stress (Mc Leman, 2009; Barnett and Webber, 2009). Yet, there is another school of thought which deliberates on capacity of social and ecological systems to adapt to the constraints, barriers of and limits to adaptation thereby giving impetus to what adaptation can achieve and highlighting that there is insufficient evidence to draw conclusion about the likelihood of migration as an adaptation strategy (Mortreux and Barnett, 2009). They emphasize on the fact that encouraging migration as a solution to climate change impacts detracts from the need for adaptation policies and allow people to 'lead the kind of life people value in places where they belong' (Adger and Barnett, 2005).

Context and issue

Migration as adaptation has been promoted in literature, but does it also resonate with ground perspectives of people who have to adapt? Do people – place bonds, local culture, place identity and community cohesion – limit this whole concept of migration as adaptation? As the United Kingdom Foresight (2011) report highlights that while environmentally motivated migration poses a challenge, the issue of people unable/unwilling to leave dangerous/risky circumstances/regions 'maybe more or equally significant'. Therefore, migration as an adaptation to climate change can be constrained by variations in risk perceptions, distant nature of climate change and failure to link current experiences with future events; and people not willing to leave their homelands. Though migration can be an adaptation strategy or a survival option, there are a number of other interacting factors which determine people's decision to move or stay. The psychological literature shows that most individuals tend to respond to issues, risks or concerns they consider as immediate and personally relevant (Moser and Dilling, 2004). According to Adger et al. (2009), climate change and its impacts, although concerning, are also generally believed to be removed in space ('not here') and time ('not yet'). Can physical impacts alone be considered as trigger for people to consider leaving their homeland, and if migrating is adaptation, then can it be considered successful if it erodes people's values, culture and place belongingness? As Adger and colleagues (2011) illustrate, there are 'limits to the idea of adaptation, too – changes such as migration may ostensibly be adaptations, but cannot be considered successful if they result in damage to people's traditions, knowledge, social orders, identities, and material cultures'. Should adaptation not focus on decreasing vulnerability, releasing migration pressures and allowing people to stay in their homeland and established communities?

This paper discusses different case studies around the world and highlights how the vision of an apocalyptic future and migration as means of adaptation is not shared by locals, who do not wish to leave and want in situ adaptation to be the focus.

When migration is adaptation

Migration has often been presented as a positive climate adaptation strategy. However, the statements presented in the literature do not clearly distinguish between conventional versus climate-induced migration. For instance, Scheffran and colleagues (2012) have described migrants as active social agents who can play an influential role in contributing to resilience and innovation in climate adaptation. They highlight how the multifaceted relationship between migration and climate adaptation interacts in three ways and that includes 'migration-for-adaptation'. The case studies presented to substantiate this statement show how migrants have initiated and implemented development projects using their financial capital and social capital in the origin and destination countries in north-west Africa. However, to what extent these development projects have the additionality factor to be categorized as adaptation is not very clear. Also, migration in that region has been a traditional strategy in response to climate extremes such as drought. Hence, the findings of the study largely refer to development benefits contributed by the migrants in their home countries and are not necessarily induced by climate change. Similarly, other scholars have supported migration as positively contributing to enhance the adaptive capacity of the communities in the vulnerable regions (Black et al., 2011; Thornton, 2011). The positive contribution, however, has been heavily focused on remittances from migrants and their positive effects such as sustaining access to basic needs in times of livelihood shocks such as drought. Better access to financial resources can definitely help in coping with external shocks, but to what extent can these be sufficient in addressing the objectives of adaptation can be explored further. McLeman and Smith (2006) have developed a conceptual model to investigate population migration as a possible adaptive response to climate change risks. Barnett and Webber (2010) have also presented migration as positively contributing to adaptation through benefits of remittances and migrants as channels of better access to knowledge and understanding about the world including climate change risks for the communities in their homeland. However, migration in this case is in search for better employment opportunities, a process that has been common traditionally. Though migration for better employment can contribute in increasing the adaptive capacity of the population, but can migration be asserted positively as adaptation to climate change. The study also clearly mentions that community resettlement should be the last resort for coping to climate change, and even in low-lying highly vulnerable islands there is a need to explore the complete range of adaptation options in the region, their barriers and limits.

Migration for strengthening the process of adaptation has been largely discussed at the background for remittances, by providing income diversification and better access to information and social networks (ADB, 2012). However, the premise here is largely development focused wherein migration contributes to alleviate poverty. On the contrary, Sundari (2005) notes that the poorest households sell their assets upon leaving and experience a loss of land, housing, jewellery, livestock and livelihoods, thereby becoming more vulnerable than they were before moving. A lack of properly planned migration can pose serious risks to migrating communities.

Migration can be beneficial in contributing towards better financial access and in a way better coping strategies in the context of climate change, but it is again place and context specific. However, literature to support the argument that migration or rather resettlement can be an adaptation strategy for the communities is limited. According to

Warner (2010), there is still a need to discuss migration systematically in the context of adaptation strategies to climate change and while some forms of environmentally induced migration may be adaptive, forced migration and displacement might be a result of the failure of socio-ecological system to adapt.

Are people migrating to adapt?

The physical impacts of climate change are visible and uncontested, but do the vulnerable communities perceive themselves as threatened and migration as a probable way of adapting to these changes? Is their risk perception of climate change in convergence with what the popular narratives emphasize? Do the dominant perspectives of them being 'victims' affect their everyday priorities, their decisions or responses to climate change? As Marino (2012) emphasizes, discourses surrounding climate change is producing additional stressors on vulnerable communities and she calls it 'insults and injuries of intervention'. Hulme further highlights a strong need for attention to the manner and language in which climate change is portrayed and translated for mass consumption. There is a tendency to dilute how societies respond to climate change as a simple cause – effect connection, wherein the inherent working structure of society is ignored and their risk perception, their response, their reasons for action versus inaction and their priorities and needs for adaptation are discounted. In climate change literature, migration has often been proposed as an adaptation strategy; but recently, many context-specific case studies have questioned this common understanding.

McNamara and Gibson (2009), while recording perceptions of Pacific Island ambassadors, assert that the focus of these nations is centred on retaining territory, nationality and cultural identity. They oppose the 'exodus' scenario imagined by the outsiders labelling them as 'climate refugees' in waiting. Instead, they appeal for assistance and resources which can allow them to live in their homeland.

Sometimes, populist and media construction of climate risk and how people respond to those risks can lead to a wrong understanding of the situation. Emerging empirical evidence contests that migration is a considered adaptation strategy by the people in question. In the Pacific Islands, the island communities have often been projected as those in crisis and likely to be future climate change refugees (Lazarus, 2009). However, mobility has been a part of the islanders past and present. How do they perceive this new alarmist manifestation of themselves? For example, in Tuvalu, Mortreux and Barnett (2009) and Farbotko and Lazarus (2012) highlight how Tuvaluans are popularly projected as first climate refugees. However, the local people fail to agree with their media constructed status. The Tuvaluans view neither their island as imminently disappearing nor their communities in crisis. The common picture of them invading borders as refugees is wrongly constructed and not shared by the locals who value their local lifestyle, culture and community cohesion on the island and strongly reject these populist narratives. Though they understand and acknowledge the risks from climate change, their approach to adapting to those risks is embedded in their valuation of their everyday lives of being islanders, and even if resources exist for them to migrate or relocate from the island, their prioritization, perception and valuation can lead to decisions which may seem irrational from the outside, but when understood from within are extremely rationale. Lazarus (2009) further notes that migration is a part of Tuvaluans' identity and belonging; however, presenting it as an ultimate future option wherein the possibility of living in their homeland is little, and adaptation is only possible by means of migration, imposes a mistaken idea about adaptation wherein it discounts the needs of the people, their priorities and willingness to not leave and live in their homeland. People have adapted and coped with climate-related impacts by means of moving away from an area of risk, but with the possibility of returning to the place which is their home. Their traditional adaptation strategy of migrating is now being forecasted as the only solution to their survival. Adger et al. (2011) points to the need for policies and interventions that enable people to adapt in ways that allows them to lead the kind of lives they value in places they call home rather than simply to foster adaptation. Similarly, in Shishmaref, Alaska, most regions are losing their land to sea erosion and relocation is the only survival option. Marino (2012) predicts that marginalized and minority communities are most likely to stay in these disaster prone areas/homeland. Vulnerability to climate change is not the only priority for the local people of Shishmaref. They too, like other regions, have their traditions, culture and lifestyle and people continue to live their lives. She further provokes by highlighting how Shishmaref is of interest to public policy and researchers only as case study of 'climate change migration', while the local communities are interested in dialogue with these policymakers and researchers as they consider that 'Shishmaref is worth saving'. Likewise in India, Upadhyay and colleagues (2015) note that the islanders from Lakshadweep do not consider climate change as a personal risk or everyday priority determining decisions to migrate. The islanders do not want to leave their home (islands). The authors argue that a sense of 'place' shapes the identity, local culture and belief systems and also contributes to their endemic resilience. Proposing migration here can erode these everyday values and resilience of the people, thereby rendering them more vulnerable than before. Should adaptation then not focus on these priorities and needs of people and should the basic right to live in their homeland be protected? Marino (2012) further illustrates the importance of incorporating local voice, priorities and concerns while relocating people from vulnerable regions, while also considering their social histories and traditional responses to natural hazards in order to facilitate relocation.

Maldives, the island nation made famous by the 'come before they sink' reference and the historic underwater meeting led by its president in 2009, has made it a sought-after destination for media stories on climate refugees and migration. As early as 1998, the then President Maumoon Abdul Gayoom expressed resistance to imposition of them as climate refugees with the vision of an apocalyptic future. He insisted on respecting them as sovereign citizens who do not wish to leave their homeland and urged for international support and action to help them stay (Gayoom, 1998). Though climate change and its impacts are uncontested in these islands, the push to publicize numbers of migrants may not be shared by the local people who do not want to leave their island and communities. Also, though the islanders observe changes in their island, it is challenging to ascertain what a certain level of sea level rise means when it is translated into daily lives. President Gayoom appealed for the international community to mobilize attention and resources to assist them live in their homeland. He insisted on sovereignty of his people, in the face of media constructions and policy discourse produced by outsiders who suggested relocation as the only option. Instead of debating migration numbers, he asserted the focus should be on reducing climate change impacts like sea water intrusion, degrading costal soil fertility which would affect the day-to-day lives of the islanders. McNamara and Gibson (2009), while interviewing Pacific Island ambassadors, reported a common argument which was stressed by most that the 'focus on migration rather than mitigation was not only defeatist but also globally irresponsible vision of the future'.

India is home to many migrants. Migration across India's borders (from the neighbouring countries) has been for a broad range of reasons such as economic, political, socio-cultural and historical linkages. Flow of migration has taken place generally from Bangladesh, Nepal, Sri Lanka, Tibet, Myanmar and Pakistan. It is interesting to note that environmental causes or climate change does not categorize as one the drivers of migration in the Census of India data. The key drivers for migration are 'employment', 'business', 'education', 'marriage' and 'Others'. Other drivers such as natural disasters, social/political problems, housing problems and migration of parents are combined in 'Others'. For climate change–induced migration, any numbers will be extracted from 'others' category, thereby giving highly doubtful numbers. While discussing climate change and migration in the Indian context, movement from Bangladesh to India is an immediate link that comes to mind. But this cross-border migration has been substantial and considered a time-established process with maximum people moving to the Indian states of West Bengal and Assam (Samaddar, 1999; Ramachandran, 2005; ADB, 2012). Likelihood of increased economic opportunities, cultural and lingual similarities are the prime motivators for this kind of movement (Samaddar, 1999; Alam, 2003; Lahiri-Dutt, 2004; ADB, 2012). Several researchers have argued that climate change will play a major role in inducing people to migrate from Bangladesh to India on a large scale (Rajan, 2008; Panda, 2009). Myers (2002) adds that climate refugees from Bangladesh alone might outnumber all current refugees worldwide. He projected that 26 million refugees will come from Bangladesh. According to Homer Dixon (1994), Bangladeshi migrants have expanded the population of India by 12–17 million over the last 40 years caused by environmental scarcity. But these claims are supported with little empirical or ground evidence, wherein climate change can be identified as a primary driver. This unexplained crunching of numbers raises questions on how these numbers are calculated and even more serious concerns over their attribution to climate change. Migration is a complex interplay of multiple factors (Perch-Nielsen et al., 2008). Bates (2002) opines that environmental changes affect migration decisions only after being filtered through the local socio-economic context. Climate change may emerge as new driver through a multiplier effect on existing push and pull factors of migration; however, a lot of uncertainty remains in assigning climate change as a significant reason. In this situation, where migration is already an established process, assigning migration as probable climate adaptation strategy raises two main thoughts for discussion: (1) if migration is accepted as adaptation for people moving from Bangladesh to India, then can it be considered successful if it leads to communal and resource conflicts in India; and (2) should the focus not be on discussing resources and finance for Bangladesh, which is a renowned pioneer in community-based adaptation rather than promoting migration as an adaptation strategy. Would not migration, if labelled as adaptation in this case, discount the very concept of adapting?

In 2005, the UNEP predicted that climate change would create 50 million climate refugees by 2010. It was anticipated that people would flee due to a range of disasters, impacts like sea level rise, increase in the numbers and severity of hurricanes, disruption to food production and so on. However, according to Aktins (2011), these regions identified as future hotspots of climate refugees are not only *not* losing people, but they are also actually among the fastest-growing regions in the world. Questioning UNEPs forecast, in 2011, Spiegel International released a new media story entitled 'Feared Migration Hasn't Happened, UN Embarrassed by Forecast on Climate Refugees' (Bojanowski, 2011). The map was withdrawn by the UNEP citing clarification that the map was originally produced by a newspaper (UNEP, 2013). The United Nations University (UNU) responded, claiming that forecast figures of environmental migrants varied widely because researchers were unsure of climate change scenarios them-selves, unsure of how climate change would contribute to hazards and unsure about how these hazards would affect people on the ground (Fisher, 2010). This again highlights how numbers have been the focus of the climate migration debate, and the process itself and people in question, their priorities and choices have been ignored.

Discussion

Adaptation to climate change is inevitably place and context specific and if migration is promoted as an all-purpose adaptation option in every context and place, then the objectives of adaptation are dis-counted. Adger and Barnett (2011) argue a need for more geographi-cally and culturally nuanced risk appraisals that allow policymakers to recognize the diverse array of climate risks to places and cultures as well as to countries and economies. Similarly, Hess and colleagues (2008) emphasize on the importance of place for promoting resilience because identity and sense of place are central to community resilience, public health and well-being more generally.

There is little doubt that migration decisions are complex and con-text specific in nature. Climate change can act as a multiple stressor on already existing vulnerabilities and emerge as an additional driver for already existing migration behaviour. With the exception of when a person's life is directly threatened, the decision to migrate is often made because of a variety of 'push' and 'pull' factors. Rarely is the deci-sion to migrate made due to a single reason. Climate change emerges as the new driver for forced and voluntary migration, through either changing existing trends (influencing poverty, increasing competition for natural resources) or creating new ones (e.g. rapid sea level rise). A decision to move or stay is specific to people (household), societies and environment systems. Instead of focusing on the questions of 'when, where and how many migrants', there is a need to understand the dichotomy of the situation wherein:

1 If people are unwilling to move from risky regions, then what should adaptation do to enable them to live their lives in places they call home?
2 Understanding how communities become pushed towards migration and what are the challenges for successful relocation.

Migration as adaptation maybe a grey area, but climate-induced movement raises serious questions of human rights, sovereignty and concerns of place identity, culture, values, psychological health and pressure on host community amongst others. Understanding local perspectives on these issues, prioritizing their adaptation needs and making them a primary stakeholder during the conception, design and implementation process are essential for successful adaptation.

While the question of survival can force people to migrate with little choice to stay, proposing migration without careful consideration of impacts pre- and post-migration can lead to chaos from legal, political, social and cultural point of view. While proposing migration as a probable adaptation option, assessment of social impacts like loss of values, culture, community cohesion and varied risk perceptions is important. There needs to be careful consideration of place identity and inclusion of person–place bonds to facilitate planning and policy development that is appropriate to local context (Baxter and Armitage, 2012). In Kiribati, Kuruppu (2009) highlights how people are unwilling to migrate due to sea level rise and discusses the role of religion for adaptation and how it can facilitate adaptation. Understanding vulnerabilities and needs of adaptation of specific people, societies and their systems and how they perceive climate risks and what are their attitudes towards them adapting to this change is thus critical to successful adaptation. As noted earlier, in situ adaptations will be the most common response to climate change. Policymaking, therefore, needs to address what is needed to allow communities to live in places they call home and aim to strengthen their preparedness and resilience. As migration may ostensibly be adaptation, it cannot be considered successful if it damages people's traditions, knowledge, social orders, identities and material cultures (Adger and Barnett, 2011).

Last, people living in risky and vulnerable regions might be unable to link current experiences of climate change with future events. The distant nature of climate change may lead them to act myopically, discounting the future risk. While moving to areas of less risk by means of migration would be the choice emerging more strongly in the future, their current actions and responses could be rooted in the present and visible climate impacts. When climate change is not an everyday priority, the whole concept of migrating due to climate change even in the categorized vulnerable regions will not follow the popularized trends. In many cases, migration will be the last resort, which will only be considered if the different adaptation strategies have failed. Therefore, though migration could be a strategy to move away from an area of risk, it should not overshadow the objectives of adaptation.

The objective of this paper is not to reject migration as a climate adaptation strategy but rather to initiate discussion on people's priorities and need for innovative and context-specific adaptation strategies, which gives them a possibility to stay in their homeland rather than being forced to migrate.

The way forward

Identifying the needs of adaptation in regions where migration is anticipated by outsiders but not shared by local communities, who are unwilling to leave, is critical to ensure that people can be better prepared and continue to live in their homeland. Migration is a complex, highly subjective and context-specific process, and climate change will only add to this challenge. However, public policy needs to deliberate on the possibilities and be prepared to manage these kinds of movements.

1 Migration is a very context-specific process, and therefore geographically and culturally nuanced assessments at the ground level are a prerequisite before arriving at any conclusion about the migrants and their priorities. This kind of assessment will also help decision makers to recognize the diversity of climate risks and responses in different places and cultures. Governments and policymakers need to consciously invest in this kind of research and assessments in order to better understand ground situations and thereby make informed decisions which are both politically and culturally acceptable.

2 The discourse in the international negotiations should not be limited to relocation of vulnerable communities and discussing

migrant numbers, but rather focus on in situ adaptations which facilitate people to adapt while still living in their home countries. Accordingly, identification and prioritization of local adaptation strategies is required, which can help in building resilience by minimizing the impacts on natural resources and livelihoods of communities.

3   Loss and damage work is an emerging debate in international negotiations in the context of climate adaptation, but with little clarity about what constitutes loss and what damage is. In the context of climate-induced migration, there is likely to be erosion of local culture, place, values and traditional knowledge. As these cannot be easily monetized or assigned quantitative values, international policy needs to discuss how this kind of loss from climate change can be estimated and how resources will be allocated for these elements which cannot be reduced to economic metrics. On the international platform, there is a need for constructive debate to discuss how to compensate loss of homeland, culture and values. Policy needs to deliberate on what will be the criteria for distribution of resources, wherein it is challenging to establish what the loss is and how much is the damage. All these questions need to be discussed carefully at the international platform.

4   Climate change and migration is a multilayered and dynamic process which is still not completely understood. Policymaking in such a context, where climate change as a cause of migration is uncertain and highly contextual and people specific, necessitates inclusivity of local populations in decision-making, wherein their views/perceptions and responses are democratically represented and not merely channelled into participatory programme processes. Their inclusivity needs to be built in from the conception, design to implementation of any policy developed to address this complex issue.

5   Data scarcity often plagues the empirical explanation of climate change and migration links. This leads to creative methods for estimating the magnitude of past, current and future climate change–induced migration – methods that are generally controversial. This lack of adequate data, particularly in terms of time series of environmental and demographic variables, is a constraint for methodological innovation and any conclusive results. In order for any analysis or empirical explanation to assist decision-making, policy needs to invest in data collection and management. Data collection agencies need to be sensitized about climate change as plausible driver for migration amongst already existing drivers.

Institutional capacities need to be strengthened for understanding these issues so that when data is collected on the ground, there is a background and knowledge about this issue.

Acknowledgements

The authors gratefully acknowledge the financial support received from UNESCO, India, for the development of this paper. Further, the authors would also like to thank The Energy and Resources Institute (Teri), India, where this paper was conceived.

References

ADB (Asian Development Bank). 2012. *Addressing Climate Change and Migration in Asia and the Pacific, Mandaluyong City.* Philippines: Asian Development Bank.

Adger, N. and Barnett, J. 2005. Compensation for climate change must meet needs. *Nature*, 435, p. 328.

Adger, W. N., Barnett, J., Chapin III, F. S. and Ellemor, H. 2011. This must be the place: Underrepresentation of identity and meaning in climate change decision-making. *Global Environmental Politics*, 11, pp. 1–25.

Adger, Neil W., Katrina Brown, Donald R. Nelson, Fikret Berkes, Hallie Eakin, Carl Folke, Kathleen Galvin, Lance Gunderson, Marisa Goulden, Karen O'Brien, Jack Ruitenbeek and Emma L. Tompkins. 2011. Resilience implications of policy responses to climate change. *Wiley Interdisciplinary Reviews: Climate Change*, 2(5), pp. 757–766.

Adger, Neil W., Irene Lorenzoni and Karen L. O'Brien. 2009. *Adapting to Climate Change: Thresholds, Values and Governance.* Cambridge: Cambridge University Press.

Agrawal, Arun and Nicolas Perrin. 2009. Climate adaptation, local institutions and rural livelihoods. In W. Neil Adger, Irene Lorenzoni and Karen L. O'Brien, eds, *Adapting to Climate Change: Thresholds, Values and Governance.* Cambridge: Cambridge University Press. Chapter 22, pp. 350–367.

Aktins, G. 2011. What Happened to the Climate Refugees? 11 April 2011. asiancorrespondent.com. [Online]. Available at: http://asiancorrespondent. com/52189/what-happened-to-the-climate-refugees/ [Accessed 24 February].

Alam, S. 2003. Environmentally induced migration from Bangladesh to India. *Strategic Analysis*, 27(3), pp. 422–438.

Barnett, J. and Webber, M. 2009. Accommodating Migration to Promote Adaptation to Climate Change, Swedish Commission on Climate Change and Development, Stockholm.

Barnett, J. and Webber, M. 2010. Background Paper to the 2010 World Development Report. Accommodating Migration to Promote Adaptation to Climate Change. The World Bank, 62pp.

Bates, D. C. 2002. Environmental refugees? Classifying human migrations caused by environmental change. *Population and Environment*, 23(5), pp. 465–477.

Baxter, J. A. and Armitage, D. 2012. Place identity and climate change adaptation: A synthesis and framework for understanding. *WIREs Climate Change*, 3, pp. 251–266. doi: 10.1002/wcc.164.

Bhagat, S. 2009. What about 30 million climate refugees? *The Times of India*, 15 December 2009. [Online]. Available at: http://timesofindia.indiatimes.com/world/europe/What-about-30-million-climate-refugees/articleshow/5338176.cms?referral=PM [Accessed 24 February 2014].

Black, R., Bennett, S. R. G., Thomas, S. M. and Beddington, J. R. 2011. Migration as adaptation. *Nature*, 478, pp. 447–449.

Bojanowski, A. 2011. Feared Migration Hasn't Happened: UN Embarrassed by Forecast on Climate Refugees. Spiegel Online, 18 April 2011. [Online]. Available at: www.spiegel.de/international/world/feared-migration-hasn-t-happened-un-embarrassed-by-forecast-on-climate-refugees-a-757713.html [Accessed 24 February 2014].

Brown, O. 2008. *Migration and Climate Change in Africa: An Overview*, IOM Migration Research Series 31. Geneva: International Organization for Migration.

Bulman, E. 2005. *U. N. Warns of Millions of Environmental Refugees*. The Associated Press, 12 October 2005. [Online]. Available at: http://seattletimes.com/html/nationworld/2002555243_refugees12.html [Accessed 24 February 2014].

Castles, Stephen. 2002. Migration and Community Formation Under Conditions of Globalization. *International Migration Review*, 36(4), pp. 1143–1168.

Census of India. 2001. Accessed at: http://censusindia.gov.in/Census_And_You/migrations.aspx?drpQuick=&drpQuickSelect=&q=migration+2011 [Accessed 22 April 2014].

Christian Aid. 2007. *Human Tide: The Real Migration Crisis*. London: Christian Aid .

Farbotko, C. and Lazarus, H. 2012. The first climate refugees? Contesting global narratives of climate change in Tuvalu. *Global Environmental Change*, 22, pp. 382–390.

Fisher, D R. 2010. COP15 in Copenhagen: How the merging of movements left civil society out in the cold. *Global Environmental Politics*, 10(2), pp. 11–18.

Foresight: Migration and Global Environmental Change. (2011). Final Project Report, the Government Office for Science, London Gayoom, M. A., 1998. International assistance can save our peoples. In M. A. Gayoom, ed., *The Maldives: A Nation in Peril*. Malé: Ministry of Planning Human Resources and Environment, p. 29.

Gayoom, M. A, 1998. International assistance can save our peoples. In M. A. Gayoom, ed., *The Maldives: A Nation in Peril*. Malé: Ministry of Planning Human Resources and Environment, p. 29.

Hartmann, B. 2010. Rethinking climate refugees and climate conflict: Rhetoric, reality and the politics of policy discourse. *Journal of International Development*, 22(2), pp. 233–246.

Hess, J. J., Malilay, J. N. and Parkinson, A. J. 2008. Climate change: The importance of place. *American Journal of Preventive Medicine*, 35(5), pp. 468–478.

Homer-Dixton, Thomas, F. 2009. Environmental scarcities and violent conflict: Evidence from cases. *International Security*, 19(1), pp. 5–40.

Hulme, M. 2008. Geographical work at the boundaries of climate change. *Transactions of the Institute of British Geographers*, 33, pp. 5–11.

Intergovernmental Panel on Climate Change (IPCC). 1990. *First Assessment Report* [W. J. McG. Tegart, G. W. Sheldon and D. C. Griffiths, eds]. Canberra, Australia: Australian Government Publishing Service.

International Organization for Migration (IOM). 2008. *Climate Change and Migration: Improving Methodologies to Estimate Flows, No. 31*. Switzerland: International Organization for Migration, Research and Publications Unit.

Kuruppu, N. 2009. Adapting water resources to climate change in Kiribati: The importance of cultural values and meanings. *Environment Science and Policy*, 12, pp. 799–809.

Lahiri-Dutt, K. 2004. Fleeting land, fleeting people: Bangladeshi women in a charland environment in Lower Bengal, India. *Asian and Pacific Migration Journal*, 13(4), pp. 475–495.

Lambert, J. 2002. *Refugees and the Environment: The Forgotten Element of Sustainability*. Brussels, Belgium: The Greens/European Free Alliance in the European Parliament.

Lazarus, J. Richard. 2009. Super wicked problems and climate change: Restraining the present to liberate the future. *Cornell Law Review*, 94, pp. 1153–1234.

Marino, E. 2012. Editorial-special issue introduction: Adding insult to injury: Climate change and the inequities of climate intervention. *Global Environmental Change*, 22, pp. 323–328.

McLeman, R. 2009. Climate change and adaptive human migration: Lesson from rural North America. In W. N. Adger, ed., *Adapting to Climate Change: Thresholds and Governance*. Cambridge, UK: Cambridge University Press, Ch.19, pp. 296–310.

McLeman, R. A. and Smith, B. 2006. Migration as an adaptation to climate change. *Climatic Change*, 76(1), pp. 31–53.

McNamara, Karen Elizabeth and Gibson, Chris. 2009. "We do not want to leave our land": Pacific Ambassadors at the United Nations Resist the Category of "Climate Refugees". *Geoforum*, 40(3), pp. 475–483.

Mortreux, C., and Barnett, J. 2009. Climate change, migration and adaptation in Funafuti, Tuvalu. *Global Environmental Change*, 19, pp. 105–112.

Moser, S. C. and Dilling, L. 2004. Making climate hot: Communicating the urgency and challenge of global climate change. *Environment*, 46(10), pp. 32–46.

Myers, N. 2002. Environmental refugees: A growing phenomenon of the 21st century. *Philosophical Transactions of the Royal Society*, 357(1420), pp. 609–613.

Myers, N. 2005. Environmental refugees: An emergent security issue. In: Session III – Environment and Migration, 13th Meeting of the OSCE Economic Forum, Prague, Czech Republic, 23–27 May 2005, pp. 23–27.

Panda, A. 2010. Climate refugees: Implications for India. *Economic and Political Weekly*, 14(20), pp. 76–79.

Perch-Nielsen, S. L., Battig, M. B. and Imboden, D. 2008. Exploring the link between climate change and migration. *Climatic Change*, 91, pp. 375–393.

Rajan, C. S. 2008. *Climate Migrants in South Asia: Estimates and Solutions.* Bangalore: Greenpeace India Society.

Ramachandran, S. 2005. *Indifference, Impotence, and Intolerance: Transnational Bangladeshis in India*, Global Migration Perspectives, No. 42. Geneva: Global Commission on International Migration.

Samaddar, R. 1999. *The Marginal Nation: Transborder Migration from Bangladesh to West Bengal.* New Delhi: Sage.

Scheffran, J., Marmer, E. and Sow, P. 2012. Migration as a contribution to resilience and innovation in climate adaptation: Social networks and co-development in Northwest Africa. *Applied Geography*, 33, pp. 119–127.

Sherriff, L. 2005. Environmental Refugees Could Hit 50m by 2010. The Register. [Online]. Available at: www.theregister.co.uk/2005/10/12/environmental_refugees/ [Accessed 24 February 2014].

Stern, N. 2006. *The Stern Review on the Economic Effects of Climate Change (Report to the British Government).* London: HM Treasury.

Sundari, S. 2005. Migration as a livelihood strategy: A gender perspective. *Economic and Political Weekly,* 40(22–23), pp. 2295–2303.

Thornton, F. 2011. Regional labour migration as adaptation to climate change: Options in the pacific. In M. Leighton, X. Shen, and K. Warner, eds, *Climate Change and Migration: Rethinking Policies for Adaptation and Disaster Risk Reduction.* Germany: United Nations University. Chapter 2.1, pp. 81–89.

UNEP. 2013. Environmentally Induced Migration Map – Clarification. [Online]. Available at: www.grida.no/general/4700.aspx [Accessed 24 February 2014].

UNFCCC. 2014. Database on Local Coping Strategies. [Online]. Available at: http://maindb.unfccc.int/public/adaptation/ [Accessed 24 February 2014].

Upadhyay, H., Kelman, I. and Mohan, D. 2015. Everyone Likes It Here. Forced Migration Review, Issue 49. [Online]. Available at: www.fmreview.org/climatechange-disasters/upadhyay-mohan-kelman [Accessed 24 September 2015].

Warner, K. 2010. Global environmental change and migration: Governance challenges. *Global Environmental Change*, 20, pp. 402–413.

4 Exploring vulnerability in flood-affected remittance-recipient and non-recipient households of upper Assam in India

Soumyadeep Banerjee, Dominic Kniveton, Richard Black and Partha Jyoti Das

The overwhelming focus on causal linkages between environmental stressors and the decision to migrate, disagreement among stakeholders regarding the positioning of migration within climate change discourses and the lack of empirical evidence about the role of migration in climate change adaptation (CCA) have been major impediments to mainstreaming of migration within adaptation policies and programmes. There has been a rise in the publications that refer to migration as an adaptation during the last decade (see McLeman and Smit 2006; Tacoli 2009; Foresight 2011; ADB 2012; Warner et al. 2012; Gemenne and Blocher 2016). The Cancun Adaptation Framework, which was signed at COP 16 in 2010, formally considered migration to be a form of adaptation to climate change by the UNFCCC signatories (McLeman 2016). The inclusion of paragraph 14f in the UNFCCC text has provided an opportunity to mainstream migration into national adaptation plans (UNFCCC 2011). However, this issue has received little attention within the national climate change discourse across the Hindu Kush Himalayan region. The public policy in this region perceives migration as a challenge to development and adaptation goals. The ADB (2012) suggests that the scattered nature and inadequacy of policy responses and normative frameworks that address climate-induced migration is due to the lack of reliable data about the nature and extent of population movements (including those related to environmental changes), limited comprehension of the nature of migration and little attention received by climate change and migration relationship from public policy until recently. The extent to which migration can contribute to CCA among migrant-sending households and origin communities is a complex issue and requires further exploration.

The IPCC's AR5 defines adaptation as 'the process of adjustment to actual or expected climate and its effects. In human systems, adaptation seeks to moderate or avoid harm or exploit beneficial opportunities' (IPCC 2014, p. 5). A key component of adaptation is the reduction of vulnerability of a system to climate change and variability. In order to understand the complex issues that may arise in different communities due to impacts of climate change, there is a need to situate the environmental stressors and shocks (including climate variability) within pre-existing scenario in specific places at specific times that had been shaped by human societies, social hier-archies, economic marginalisation, entitlements, institutional capa-bilities and political systems (Bohle et al. 1994; Hahn et al. 2009; Shah et al. 2013). The susceptibility of physical and social systems to be harmed can be described through the concept of vulnerability (Adger 2006). The analysis in this chapter adopts the IPCC con-ceptualisation of vulnerability as a function of three major compo-nents, viz. sensitivity, exposure and adaptive capacity. The context is critical to vulnerability. The characteristics of system, the type of hazard, the region, the population group and the time period will shape the factors that make a system vulnerable to a hazard (Down-ing and Patwardhan 2004; Brooks and Adger 2005). Within the climate change research, diverse methodologies have been used to assess vulnerability of different systems using primary or secondary data (see Brooks et al. 2005; Vincent 2007; Hahn et al. 2009; Gupta et al. 2010; Hinkel et al. 2010).

The extent to which remittances would have a positive or negative role in remittance-recipient households is context specific (Barnett and Webber 2009; De Haas 2012). In practice, relatively little is known about the specific role of remittances in the process of reducing vulnerability to climate-related stressors. This chapter aims to understand how the choices on remittance usage already made by a household shape the vulnerability of remittance-recipient and non-recipient households to floods in upper Assam in the Eastern Brahmaputra sub-basin (EBSB) of India. This chapter is organised as follows. The next section provides an overview of the research methodology. This is followed by empirical evidence on the characterisation of vulnerability in the remittance-recipient and non-recipient households. The last section discusses the policy implications of these findings.

Research methodology

Methodology

Since the impact data is often unavailable in developing nations, the indicator-based approaches have been used in such circumstances (Adger et al. 2004), which are comparatively less costly and time consuming and could be applied to the micro- and meso-scales (Nair et al. 2013). This approach provides a framework to characterise vulnerability of a system, helps to standardise measurement and permits a comparison between different groups within a system (e.g. remittance-recipient and non-recipient households). It has been suggested by the New Economics of Labour Migration that migration is a household-level strategy (Stark and Bloom 1985). The present research envisages that the household, which occupies a specific geographical location, could be connected to multiple locations through a migrant worker and/or access to remittances.

The IPCC's AR5 defines vulnerability as '[t]he propensity or predisposition to be adversely affected. Vulnerability encompasses a variety of concepts including sensitivity, susceptibility to harm and lack of capacity to cope and adapt' (IPCC Glossary 2014, p. 28). Vulnerability is a function of three major components, namely sensitivity, exposure and adaptive capacity.[1] Based on the livelihoods vulnerability literature (e.g. Vincent 2007; Eakin and Bojorquez-Tapia 2008; Hahn et al. 2009; Gerlitz et al. 2016), the sensitivity of rural households to floods in upper Assam is conceptualised to include five sub-dimensions, namely environmental dependence, water, food, well-being and health. Each of these sub-dimensions are characterised by relevant attributes that are in turn comprised of generic and specific indicators. These indicators have been identified during the focus group discussions in the study area. This is further supplemented by literature review and inputs from local experts. These indicators are organised into attributes, sub-dimensions and major components based on the conceptual framework. For example, attributes such as the dependence on crop income, crop diversification index, non-farm income diversification index, reduction in agricultural assets due to floods, quality of exterior wall of the dwelling and dependence on natural resources for the primary source of cooking fuel constitute the environmental dependence sub-dimension. The water sub-dimension is comprised of access to drinking water, access to drinking water storage, structural changes in water source to address flood impacts and access to safe drinking water during floods. The food sub-dimension is comprised of attributes associated with flood period such as reliance on less preferred food items, restriction of food consumption among adult household members, use of savings to procure food, collection of wild food (e.g. fruits, vegetables and herbs) and begging for food. The well-being of a household is represented by reduction in expenditure on education and clothes due to floods and selling or mortgaging of household assets in response to flood impacts. The health sub-dimension is represented by reduction in health spending due to floods. A detailed discussion about the conceptual framework and selection of these attributes can be found in Banerjee (2017).

United Nations (cited in Fussel 2007, p. 154) defines a hazard as 'a potentially damaging physical event, phenomenon or human activity that may cause the loss of life or injury, property damage, social and economic disruption or environmental degradation'. It will be critical to have the capacity to adjust to the frequency and severity of familiar recurrent hazards and to support systems so that they can adapt to the altered levels of hazard (Brooks and Adger 2005). The knowledge of actions surrounding past stress events (e.g. droughts, floods) has been used as a proxy to understand how systems might build and mobilise (or not) their adaptive capacity to prepare for and respond to future climate change (Engle et al. 2011). This study explores the period from 1984 to 2013, a 30-year period since the average weather for 30 years is climate. The exposure of a household to floods is comprised of three sub-dimensions: number of years between 1984 and 2013, when the household had experienced a flood; damages to the household in monetary terms during each episode of flood between 1984 and 2013; and time taken by a household to recover from the damages caused by each flood between 1984 and 2013.

Based on the adaptive capacity literature (e.g. Eakin et al. 2011; Aulong et al. 2012; Gerlitz et al. 2016) and the Sustainable Livelihoods Framework (SLF), the household-level adaptive capacity is conceptualised to be comprised of five sub-dimensions: financial assets, natural assets, social assets, human assets and physical assets. Financial assets are represented by access to formal financial institutions and insurance. The farm size diversification index, livestock diversification index, and changes in agricultural practices in response to flood are selected as attributes of a household's natural assets. The sub-dimension on social assets is comprised of access to flood assistance, and access to alternative livelihood opportunities in the locality or nearby areas are attributes of human assets. Physical assets are represented by structural changes in houses to address flood impacts and access to storage, farm mechanisation and access to transportation during floods. A discussion about the selection of these attributes can be found in Banerjee and colleagues (2017).

Based on Hahn and colleagues (2009), this study has adopted the equal weighted design to construct the vulnerability index. The major components, which are comprised of different number of sub-dimensions, contribute equally to the vulnerability index. Since the attributes are measured on different scales, each attribute is standardised. The value of a sub-dimension is estimated by averaging the attributes that constitute it. An average of sub-dimensions provides the value of their respective major components. The three major components were combined using the following equation:

$$VI_s = (EI - AI) * SI$$

Where VI_s is the vulnerability index for a household in a particular study area s, and EI, AI and SI are the exposure index, adaptive capacity index and sensitivity index, respectively, for the same household. The VI ranges from −1 to +1. An overview of estimation of vulnerability index could be found in Banerjee and colleagues (2016).

Description of the study area

The region of upper Assam, which is located in the EBSB, experiences floods on a regular basis. The heavy rainfall within a short time from June onwards due to the southwest tropical monsoon contributes to the flood risk. The physiography of the Brahmaputra basin, rise in population in flood-prone areas, the construction of new infrastructure and housing, expansion of economic activities, changes in land use, encroachment of wetland and low-lying areas, temporary flood control measures and poor maintenance of embankments contribute to drainage congestion and frequent occurrences of floods in this region (TERI 2011). The study area includes districts of Dhemaji, Dibrugarh, Lakhimpur and Tinsukia. These floods have direct and indirect effects on the lives of people in these areas. Houses are inundated by flood water. In severely affected villages, the household members have to shift to safe locations (e.g. road, embankment and relief camp). Inundation and/or damage to infrastructure disrupt transportation. The high reliance on natural resource–based livelihoods and location in a flood-prone river basin exposes the local population to flood risks. Most vulnerable among the crops is the winter or *sali* paddy, the main kharif (monsoon) crop (Mandal 2010). Flood deposits sand ('sandcasting') and other sediments that bury standing crops or render farmland unsuitable for farming (Das et al. 2009). In the aftermath of floods, crops and livestock diseases have been reported. The flood responses undertaken by the state government of Assam include setting up of relief camps, providing relief materials and compensation to the affected families, construction and maintenance of embankments and drainage channels, anti-erosion and protection measures, restoration of communication and transport infrastructure and providing flood warning information (Goyari 2005). Although local households have developed a wide range of flood responses, these strategies focus on immediate flood impacts (e.g. moving family and cattle to a safe location, building rafts and contacting district administration) and help the households to cope during the flood period (Banerjee et al. 2017). Only structural changes in house (e.g. raising plinth of the house, cattle shed and granary) and drinking water source could be considered as enhancing medium- or long-term capacity in the context of flood risks (Banerjee et al. 2017).

Farming is subsistence in nature. Paddy is the principal *kharif* ('monsoon') crop. Since rainfall is scarce during *rabi* ('winter') season, crops that are less water intensive such as potato, vegetables and mustard are grown (Mandal et al. 2015). Farming is at risk due to the vagaries of weather, cost of agricultural inputs, the volatility of the crop prices and incidence of crop diseases. It is supplemented by livestock rearing, daily wage earning, salaried income, small business or remittances. Circular labour migration is an emerging livelihood strategy in this area. Predominantly, men of working age with school education migrate to urban centres in Assam or other parts of India, and are employed in manufacturing, construction and the service sectors (Banerjee et al. 2017). Remittances are commonly invested in food, healthcare, community activities, consumer goods, education and transport (Banerjee et al. 2017).

Findings

Though the vulnerability of remittance-recipient households (–0.0778) is marginally higher than that of non-recipient households (–0.0831), this finding is not statistically significant. However, some differences between these two groups of households are statistically significant at the attribute level. Table 4.1 presents an overview of the household-level

Table 4.1 Sub-dimensions and attributes of sensitivity by remittance-recipient status of the household, upper Assam, Eastern Brahmaputra sub-basin*

Sub-dimension	Non-recipient households	Recipient households	Attributes	Non-recipient households	Recipient households
Health	0.1339	0.1480	Reduced health expenditure due to flood	0.1339	0.1480
Well-being	0.2260	0.2437	Reduced educational expenditure due to flood	0.1246	0.1480
			Reduced clothes expenditure due to flood	0.2077	0.2471
			Sold or mortgaged household assets due to flood	0.3458	0.3359
Water	0.4902	0.4936	Average time to collect drinking water for a normal day	0.1477	0.1712**
			Did not store drinking water for consumption during inundation	0.7975	0.8050
			Did not filter or boil drinking water for consumption during inundation	0.4268	0.4150
			Did not raise height of the wall surrounding the well or height of the tube well in response to flood	0.5888	0.5830
Food	0.3635	0.3865	Relied on less preferred food items due to flood	0.3068	0.2992
			Restricted food consumption among adults due to flood	0.5327	0.5772
			Collected wild food due to flood	0.2321	0.2780
			Did not spend savings to buy food due to flood	0.4626	0.4556
			Begged for food due to flood	0.2835	0.3224
Environmental dependence	0.5380	0.5428	Above median income from crop sale	0.3489	0.2625**
			Crop diversification index	0.4994	0.5504*
			Non-farm income diversification index	0.3890	0.4089
			Reduction in agricultural assets due to flood	0.3645	0.3784
			Household with exterior walls made of weak construction material	0.7382	0.7722
			Dependence on environmental resources for primary source of cooking fuel	0.8959	0.8842

* The sub-dimensions and attributes have been standardised. Legend: * $p<.1$; ** $p<.05$

Source: Computed by the author from HICAP Migration Dataset

sensitivity to flood among remittance-recipient and non-recipient households in upper Assam. Among the attributes of environmental dependence, the results for dependence on crop income and crop diversification index are significant. A household that is dependent on crop income will be sensitive to climate stressors (Adger 1999). Even though the farming in this study area is subsistence in nature, remittance-recipient households are less dependent on crop income than non-recipient households. During the year preceding the survey, the average crop income in non-recipient and remittance-recipient households was estimated to be USD 108 and USD 95, respectively. Mandal (2010) suggests that farmers in Assam have adopted crop diversification as a strategy to avoid crop losses due to frequent floods. Remittance-recipient households grew fewer crops than non-recipient households. This is indicated by the higher crop diversification index among remittance-recipient households. One of the attributes of the water sub-dimension is average time taken by a household member to collect drinking water required for a household's consumption on a normal day. A member of a remittance-recipient household took longer to accomplish this task (30.8 minutes) than a member of non-recipient household (26.6 minutes). At the sub-dimension level, the differences between remittance-recipient and non-recipient households are significant for financial and human assets (see Table 4.2).[2] Remittance-recipient households had better access to financial assets than non-recipient households. Although remittance-recipient households had better access to formal financial institutions and insurance, these differences were not significant at the attribute level. The insurance penetration remained low among these households. Only one-third of the households reported to have a life insurance policy for a household member. None had reported to have crop or livestock insurance. In context of human assets, remittance-recipient households had access to more communication devices and alternative livelihood opportunities in the locality or nearby areas than non-recipient households. Among the attributes of natural assets, the farm size diversification index is higher among remittance-recipient households than non-recipient households. This indicates that the former have smaller farm size than the latter. Also, more remittance-recipient households did not have access to farm mechanisation than non-recipient households.

The sensitivity of remittance-recipient households to floods (0.4233) is marginally higher than non-recipient households (0.4115). The exposure of remittance-recipient households (0.2338) is marginally higher than non-recipient households (0.2240). On the other hand, remittance-recipient households have a marginally higher adaptive

Table 4.2 Sub-dimensions and attributes of adaptive capacity by remittance-recipient status of the household, upper Assam, Eastern Brahmaputra sub-basin*

Sub-dimension	Non-recipient households	Recipient households	Attributes	Non-recipient households	Recipient households
Financial assets	0.5015	0.4401**	Did not have access to formal financial institution	0.3028	0.2510
			Did not have an insurance	0.6916	0.6293
Natural assets	0.4721	0.4735	Farm size diversification index	0.6498	0.6859*
			Livestock diversification index	0.2498	0.2351
			Did not make changes in agricultural practices in response to flood	0.5420	0.5174
Social assets	0.3240	0.3192	Did not have access to flood assistance	0.0934	0.1081
			Did not have access to financial borrowing during floods	0.6542	0.5946
			Did not participate in collective action on flood relief, recovery or preparedness	0.2243	0.2548
Human assets	0.6051	0.5604**	Communication device diversification index	0.4687	0.4452*
			Did not have access to alternative livelihoods opportunity in the locality or nearby areas	0.7414	0.6757*
Physical assets	0.4143	0.4353	Did not make structural changes in the house due to flood	0.1994	0.1853
			Did not mechanise farming to address flood impacts	0.6106	0.6988**
			Did not have access to boats or rafts during flood	0.1776	0.1776
			Did not have access to storage options during flood	0.6698	0.6795

* The sub-dimensions and attributes have been standardised. Legend: * $p<.1$; ** $p<.05$

Source: Computed by the author from HICAP Migration Dataset

capacity (0.4770) than non-recipient households (0.4372). However, these differences in major components of vulnerability between remittance-recipient and non-recipient households are not significant.

Discussion

Adaptation will be critical to address livelihood security in context of changes in climatic and non-climatic conditions. Developing countries, where much of the population rely on livelihoods that are sensitive to extreme weather events, have a pressing need to build their ability to adapt to the impacts of current climate variability and future climate change. The dependency of livelihoods (e.g. agriculture, livestock rearing and fishing) in rural Assam on natural resources (Das et al. 2009) and ecosystem services is evident. For example, among the total workforce, 26 per cent were cultivators and 8 per cent were agricultural labourers (DoES 2015, p. 15). Flooding is a major hazard in Assam. ADB (2006) estimates that on an average USD 47 million in annual crop production is lost due to floods in Assam, which also affects livelihoods of nearly three million people. A huge share of the state's resources is being diverted from development programmes every year in order to undertake relief, rescue and rehabilitation of flood-affected population in Assam (Goyari 2005). Pradhan and colleagues (2012) suggest that due to their exposure to floods over a long period, communities and institutions in Assam have devised several measures to manage floods such as bamboo platforms to store valuables; mixed cultivation of different paddy varieties; erosion control with bamboo, trees and sandbags; and income diversification (Pradhan et al. 2012). However, Banerjee and colleagues (2017) report that many of the household-level flood response strategies, even though are decades old, are short term and reactive in nature. For example, over three-quarters of surveyed households had built a raft from the banana plant (*bhur*). However, recurrent floods have led to a decline in availability of banana plants in some areas, and this situation may become acute in other areas of upper Assam in the future (Banerjee 2017). In addition, the extent of village-level flood preparedness (e.g. flood contingency plan, pre-designated flood shelter for villagers or livestock and meetings on flood preparedness) is limited (Banerjee et al. 2017).

The migration narrative in Assam is mainly centred on the issues of identity, ethnic relations, citizenship and illegal immigration. Hence, the discourse on environmental change and migration in Assam has largely been concerned with the illegal immigration from Bangladesh due to natural disasters, land scarcity, land degradation and poverty (see Hazarika 1993; Suhrke 1997) and its potential socio-cultural, political and economic impacts in destination (Swain 1996; Reuveny 2008). Also, there is a lack of empirical studies on the interrelationship between migration and CCA based on either mixed method or quantitative methodology in upper Assam. This chapter presents a new narrative on climate change and migration in Assam that moves away from identity-focused and securitised discourse to a migrant-centred one. It attempts to understand the differences and similarities in vulnerability of remittance-recipient and non-recipient households to floods. Even though the right to mobility, residence and practice of any profession within the territory of India is guaranteed by the Constitution of India (GoI 1950), access to government welfare schemes is not portable beyond the boundaries of the sending state (Srivastava and Sasikumar 2003). Since the migrant workers are mainly employed in the informal sector in destination, most of them do not have access to social security benefits (e.g. pension, provident fund or insurance). This informal nature of the job exposes them to economic risks (e.g. economic downturn, sudden termination of contract). The interstate migrant workers often confront hostility from certain sections of the host population. The latter perceive the migrant workers not just as an economic threat, but also a risk to the 'local' culture, language or religion (Mahajan et al. 2008). Banerjee and colleagues (2017) find that most migrant workers from upper Assam are wage employees in the informal sector; and a major share of these migrant workers are based in other states within India.

Savings and safe remittance transfer could be enabled through the increase in access to formal banking facilities for internal migrants (UNICEF 2013). Though a large number of households had access to a formal financial institution (e.g. savings bank account), few households had any targeted savings to manage environmental risks, particularly that posed by extreme weather events such as recurrent floods.[3] An awareness campaign among rural residents in the study area, particularly among women, about the financial products and services, significance of establishing creditworthiness in a formal financial institution, benefits of precautionary savings as a strategy to manage risks and insurance products is necessary to enhance financial literacy. Though ownership of mobile phones, television and radio is widespread in the study area, remittance-recipient households are likely to have better access to information devices than the non-recipient ones. The government institutions and development partners could explore the use of these devices to disseminate information on financial inclusion and literacy, disaster risk reduction, skill building, income opportunities and government programmes.

The non-recipient households were more dependent on the environment resources than remittance-recipient households. For example, non-recipient households were more dependent on crop income and had larger farm size. Though a higher diversity of crops and better access to farm mechanisation was observed among non-recipient households, it could be inferred that the livelihoods portfolio of non-recipient households are at a higher risk than remittance-recipient households. In this study area, the daily wage labour in non-farm sector is the main source of non-farm income in the locality (Banerjee et al. 2017). An environment conducive to promotion of livelihoods diversification needs to be created in this area. This involves improvement in transport and communication infrastructure, better access to market towns, creation of storage facilities, provisions of skill training opportunities and nurturing of rural enterprises. The ancillary activities such as transportation, communication and storage could also generate income opportunities. These interventions should not aim to reduce migration. Rather, the focus could be on maximisation of benefits from a household's income and reduction of climate and non-climate risks on its livelihoods portfolio.

Acknowledgement

The authors express their gratitude to Prof. Ram B. Bhagat and Prof. Irudaya Rajan for providing the opportunity to prepare this chapter. The authors would like to thank Dr Golam Rasul (ICIMOD), Dr Arabinda Mishra (ICIMOD) and Mr Nand Kishor Agrawal (ICIMOD) for their support and encouragement. Particular appreciation goes to Dr Bidhubhusan Mahapatra (Population Council) for his encouragement and insightful feedback. Special thanks to the Aaranyak team for their invaluable support during the fieldwork. This chapter is based on research that was supported by the Himalayan Climate Change Adaptation Programme (HICAP), which is implemented jointly by the International Centre for Integrated Mountain Development (ICIMOD), the Centre for International Climate and Environmental Research Oslo (CICERO) and Grid-Arendal in collaboration with local partners and is funded by the Ministry of Foreign Affairs, Norway, and the Swedish International Development Agency. The views and interpretations in this chapter are those of the authors and are not necessarily attributable to ICIMOD.

Parts of this chapter are drawn from Soumyadeep Banerjee's 'Understanding the Effects of Labour Migration on Vulnerability to Extreme Events in Hindu Kush Himalayas: Case Studies from Upper Assam and Baoshan County', a D.Phil. thesis prepared under the supervision of Prof. Dominic Kniveton and Prof. Richard Black at the University of Sussex.

Notes

1 The IPCC defines sensitivity as '[t]he degree to which a system or species is affected, either adversely or beneficially, by climate variability or change. The effect may be direct (e.g., a change in crop yield in response to a change in the mean, range, or variability of temperature) or indirect (e.g., damages caused by an increase in the frequency of coastal flooding due to sea-level rise)' (IPCC Glossary 2014, p. 24), and adaptive capacity 'as the ability to adjust, to take advantage of opportunities, or to cope with consequences' (IPCC Glossary 2014, p. 21).
2 In this analysis, the attributes of household-level adaptive capacity have been framed in a negative manner. For example, financial assets in Upper Assam include two attributes: 'did not have access to formal financial institution' and 'did not have access to insurance'. Hence, higher the value of a sub-dimension or attribute, lower will be the access of a household to that sub-dimension or attribute.
3 The access to formal financial institutions would have been further improved since the launch of Pradhan Mantri Jan Dhan Yojana (PMJDY), a national financial inclusion programme.

References

Adger, W. N., 1999. Social vulnerability to climate change and extremes in coastal Vietnam. *World Development*, 27(2), pp. 249–269.
Adger, W. N., 2006. Vulnerability. *Global Environmental Change*, 16(3), pp. 268–281.
Adger, W. N., Brooks, N., Kelly, M., Bentham, G., Agnew, M., Eriksen, S., 2004. New Indicators of Vulnerability and Adaptive Capacity. Tyndall Centre for Climate Change Research, Technical Report 7. University of East Anglia, United Kingdom.
Asian Development Bank, 2006. India: Preparing the North Eastern Integrated Flood and Riverbank Erosion Management Project (Assam). Technical Assistance Report, Project Number: 38412. Retrieved from Asian Development Bank website: http://www2.adb.org/Documents/TARs/IND/38412-IND-TAR.pdf. Accessed October 12, 2010.
Asian Development Bank, 2012. *Addressing Climate Change and Migration in Asia and the Pacific*. Mandaluyong City, Philippines: Asian Development Bank.
Aulong, S., Chaudhuri, B., Farnier, L., Galab, S., Guerrin, J., Himanshu, H. and Reddy, P. P., 2012. Are South Indian farmers adaptable to global

change? A case in an Andhra Pradesh catchment basin. *Regional Environmental Change*, 12(3), pp. 423–436.

Banerjee, S., 2017. Understanding the Effects of Labour Migration on Vulnerability to Extreme Events in Hindu Kush Himalayas: Case Studies from Upper Assam and Baoshan County. D.Phil. thesis submitted to the University of Sussex. Brighton: University of Sussex.

Banerjee, S., Anwar, M. Z., Gioli, G., Bisht, S., Abid, S., Habib, N., Sharma, S., Tuladhar, S., and Khan, A., 2016. An index based assessment of vulnerability to floods in the Upper Indus Sub-Basin: What role for remittances? In Milan, A., Schraven, B., Warner, K. and Cascone, N. (Eds), *Migration, Risk Management and Climate Change: Evidence and Policy Responses*. Global Migration Series 6. Dordrecht: Springer, pp. 3–23. DOI: 10.1007/978-3-319-42922-9.

Banerjee, S., Kniveton, D., Black, R., Bisht, S., Das, P. J., Mahapatra, B. and Tuladhar, S., forthcoming. Does financial remittance build household level adaptive capacity? A case study of the flood affected households of Upper Assam in India. KNOMAD Working Paper.

Barnett, J. and Webber, M., 2009. Accommodating migration to promote adaptation to climate change. The Commission on Climate Change and Development. Retrieved from www.ccdcom mission.org. Accessed May 8, 2012.

Bohle, H. G., Downing, T. E. and Watts, M. J., 1994. Climate change and social vulnerability: Toward a sociology and geography of food insecurity. *Global Environmental Change*, 4(1), pp. 37–48.

Brooks, N. and Adger, W. N., 2005. Assessing and enhancing adaptive capacity. Adaptation policy frameworks for climate change: Developing strategies, policies and measures, pp. 165–181.

Brooks, N., Adger, W. N. and Kelly, P. M., 2005. The determinants of vulnerability and adaptive capacity at the national level and the implications for adaptation. *Global Environmental Change*, 15(2), pp. 151–163.

Das, P., Chutiya, D. and Hazarika, N., 2009. Adjusting to floods on the Brahmaputra Plains, Assam, India. International Centre for Integrated Mountain Development, Nepal.

De Haas, H., 2012. The migration and development pendulum: A critical view on research and policy. *International Migration*, 50(3), pp. 8–25.

DoES, 2015. *Economic Survey: Assam 2014–2015*. Directorate of Economics and Statistics. Planning and Development Department. Guwahati: Government of Assam.

Downing, T. E. and Patwardhan, A., 2004. Assessing vulnerability for climate adaptation. In Lim, B. and Spanger-Siegfried, E. (Eds), *Adaptation Policy Frameworks for Climate Change: Developing Strategies, Policies, and Measures*. Cambridge: Cambridge University Press, Chapter 3.

Eakin, H. and Bojorquez-Tapia, L. A., 2008. Insights into the composition of household vulnerability from multicriteria decision analysis. *Global Environmental Change*, 18(1), pp. 112–127.

Eakin, H., Bojórquez-Tapia, L. A., Diaz, R. M., Castellanos, E. and Haggar, J., 2011. Adaptive capacity and social-environmental change: Theoretical and operational modeling of smallholder coffee systems response in Mesoamerican Pacific Rim. *Environmental Management*, 47(3), pp. 352–367.

Foresight, 2011. Migration and global environmental change future challenges and opportunities. Government Office for Science – Foresight, p. 234.

Gemenne, F. and Blocher, J., 2016. How can migration support adaptation? Different options to test the migration – adaptation nexus. Migration, Environment and Climate Change: Working Paper Series No. 1/2016.

Gerlitz, J. Y., Macchi, M., Brooks, N., Pandey, R., Banerjee, S. and Jha, S. K., 2016. The multidimensional livelihood vulnerability index: An instrument to measure livelihood vulnerability to change in the Hindu Kush Himalayas. *Climate and Development*, pp. 1–17.

Government of India (GoI), 1950. *Constitution of India*. New Delhi.

Goyari, P., 2005. Flood damages and sustainability of agriculture in Assam. *Economic and Political Weekly*, 40(26), pp. 2723–2729.

Gupta, J., Termeer, C., Klostermann, J., Meijerink, S., van den Brink, M., Jong, P., Nooteboom, S. and Bergsma, E., 2010. The adaptive capacity wheel: A method to assess the inherent characteristics of institutions to enable the adaptive capacity of society. *Environmental Science and Policy*, 13(6), pp. 459–471.

Hahn, M. B., Riederer, A. M. and Foster, S. O., 2009. The livelihood vulnerability index: A pragmatic approach to assessing risks from climate variability and change: A case study in Mozambique. *Global Environmental Change*, 19(1), pp. 74–88.

Hazarika, S., 1993. Bangladesh and Assam: Land pressures, migration and ethnic conflict. Occasional Paper, 3, p. 48. International Security Studies Program. Ameican Academy of Arts and Sciences. USA.

Hinkel, J., Schipper, S. and Wolf, S., 2010. *Review of Methodologies for Assessing Vulnerability*. Report submitted to the GTZ in the context of the project. Climate Change Adaptation in Rural Areas of India. Stockholm: Stockholm Environment Institute.

International Panel on Climate Change (IPCC), 2014. *WGII AR5 Glossary: Contribution of Working Group II to the Intergovernmental Panel on Climate Change Fourth Assessment Report*. United Kingdom and New York, NY, USA.

Mahajan, D., Sharma, D., Vivekanandan, J., and Talwar, S., 2008. *Climate Change Induced Migration and Its Security Implications in India's Neighbourhood*. Project Report No. 2008RS10. New Delhi: TERI.

Mandal, R., 2010. Cropping patterns and risk management in the flood plains of Assam. *Economic and Political Weekly*, 45(33), pp. 78–81.

McLeman, R., 2016. Conclusion: Migration as adaptation: Conceptual origins, recent developments, and future directions. In Milan, A., Schraven, B., Warner, K. and Cascone, N. (Eds), *Migration, Risk Management and Climate Change: Evidence and Policy Responses*. Global Migration Series 6. Dordrecht: Springer, pp. 213–229. DOI: 10.1007/978-3-319-42922-9.

McLeman, R. and Smit, B., 2006. Migration as an adaptation to climate change. *Climatic Change*, 76(1–2), pp. 31–53.

Nair, M., Ravindranath, N. H., Sharma, N., Kattumuri, R. and Munshi, M., 2013. Poverty index as a tool for adaptation intervention to climate change in northeast India. *Climate and Development*, 5(1), pp. 14–32.

Pradhan, N. S., Khadgi, V. R., Schipper, L., Kaur, N. and Geoghegan, T., 2012. Role of Policy and Institutions in Local Adaptation to Climate Change Case Studies on Responses to Too Much and ICIMOD, Kathmandu, Nepal.

Reuveny, R., 2008. Ecomigration and violent conflict: Case studies and public policy implications. *Human Ecology*, 36(1), pp. 1–13.

Shah, K. U., Dulal, H. B., Johnson, C. and Baptiste, A., 2013. Understanding livelihood vulnerability to climate change: Applying the livelihood vulnerability index in Trinidad and Tobago. *Geoforum*, 47, pp. 125–137.

Srivastava, R. and Sasikumar, S. K., 2003, June. An overview of migration in India, its impacts and key issues. Regional Conference on Migration, Development and Pro-Poor Policy Choices in Asia, pp. 22–24.

Stark, O., and Bloom, D., 1985. The new economics of labor migration. *The American Economic Review*, 75(2), pp. 173–178.

Suhrke, A., 1997. Environmental degradation, migration, and the potential for violent conflict. In Gleditsch, N. P. (Ed.), *Conflict and the Environment*. Dordrecht, The Netherlands: Springer, pp. 255–272.

Swain, A., 1996. Environmental migration and conflict dynamics: Focus on developing regions. *Third World Quarterly*, 17(5), pp. 959–974.

Tacoli, C., 2009. Crisis or adaptation? Migration and climate change in a context of high mobility. *Environment and Urbanization*, 21(2), pp. 513–525.

TERI, 2011. Assam State Action Plan on Climate Change, 2012–2017. Prepared for the department of environment and forest, government of Assam.

UNFCCC, 2011. The Cancun agreements: Outcome of the work of the ad hoc working group on long-term cooperative action under the convention. Decision 1/CP.16. In Conference of the Parties, pp. 1–31.

UNICEF, 2013. *Social Inclusion of Internal Migrants in India: Internal Migration in India Initiative.* New Delhi: UNESCO.

Vincent, K., 2007. Uncertainty in adaptive capacity and the importance of scale. *Global Environmental Change*, 17(1), pp. 12–24.

Warner, K., Afifi, T., Henry, K., Rawe, T., Smith, C. and De Sherbinin, A., 2012. *Where the Rain Falls: Climate Change, Food and Livelihood Security, and Migration.* Global Policy Report of the where the Rain Falls Project. Bonn: CARE France and UNU-EHS.

5 Institutional response to displacement due to chronic disasters

The art of muddling through

Ritumbra Manuvie

Migrations caused due to chronic hydro-meteorological disasters are a significant phenomenon in India. While there is a lack of reliable data on internal displacements, it is estimated that between 1990 and 2016 there were approximately 192 instances of floods affecting 573 million people and amounting to a damage of USD 51,746 million. Around 8 million people were left homeless and 36,000 lost their lives (EM-DAT, 2016). In accessing the performance of a government towards flood disasters, scholars and critics often employ the principals of economic and political rationality, and analyse the desirability of the strategies adopted to offset the annual flood damage in a region. In the process, the cost-effectiveness of the flood mitigation and flood management policies are scrutinized to provide a review of the policy and bring out the implementation gaps. Most researchers agree that the response to flood and flood-induced displacements in India is reactionary, inefficient, ineffective and non-reliable. The current chapter, however, questions this ontology of criticism about the governmental failure by arguing that the laggard process of policy building and response towards disasters and disaster-induced displacements is normatively desirable to achieve the required level of functionality in a divided society. Although the gradual development of the policy and the protracted implementation of interventions jeopardize the essential rights of disaster victims, from a bureaucratic perspective it provides a methodical approach to pursue welfare policies in a polarized society. This chapter extends the paradigm of incrementalism and argues that a muddled through response is an essential and humane process in which day-to-day administration can be managed. The chapter draws on analysis and interpretations from the high-level elite interviews with bureaucrats in the State of Assam and highlights how social and political constructions in a society necessitate incremental responses instead of drastic policy changes.

The remainder of this chapter is divided into three parts. The second part discusses the problem of floods and flood-induced displacement in the State of Assam. The third part examines the flood management practices in the State of Assam in the light of its political circumstance and the fourth part evaluates why the rationality criterion are non-desirable by senior bureaucrats who prefer a gradual and incremental growth in policy response.

Flood and flood displacement in Assam

The State of Assam situated in the eastern borderland of India is an environmentally sensitive region with a high nested vulnerability towards the geophysical hazards and the hydro-meteorological hazards such as earthquakes and floods (Ravindranath, et al., 2011; Chaliha, et al., 2012; Watmough, et al., 2016). There are two river systems in the state – the Brahmaputra River and the Barak River system. The state is located in a high rainfall zone receiving an average rainfall of 235 centimetres per year. Seventy per cent of this downpour is concentrated within a short span of the monsoon season starting from June until September. Historically, the monsoon season brings a massive flow of water in the river Brahmaputra and its tributaries; however, due to the change in river morphology caused by a strong earthquake in 1950, the increasing water levels of the monsoon period has caused massive floods. Over the years, the intensity of the floods has increased several times due to uneven development, land-use pressure, high population growth in the state and to some extent due to changing weather patterns (Goswami, 1998; Kotoky, et al., 2003; Singh, 2007).

The Shukla Commission Report estimates that around 31.5 lakh hectares, which constitutes 92.6 per cent of the total cultivated area of Assam, is prone to flooding and approximately 16.3 lakh hectares of this area do not even have basic physical protection for flood control (Shukla, 1997). Flood damage in the state has risen exponentially over the years and has been estimated to be INR 15.4 crore between 1970 and 1979; INR 175.3 crore between 1980 and 1989; INR 860 crore between 1990 and 1998; and a rising INR 3,618.93 crore between 2000 and 2008 (Table 5.1).

Being a rural state, 74 per cent of Assam's working population relies on agriculture and allied activities for their livelihood. Assam is known for its smallholding subsistence agriculture practices which have obstructed the use of modern farming inputs to increase crop yields. Thirty-one per cent of the population in the state lives below poverty line and lack the financial capacity needed to cope with the repeated

Table 5.1 Population affected by floods and human lives lost in Assam, 2000–2013

Year	Area affected (lakh hectares)	Crop area affected (lakh hectares)	Population affected (in million)	Human lives lost	Cattle lost	Total damage (rupees in crore)
2000	9.66	3.22	38.88	36	19,988	251.18
2001	2.03	0.36	5.42	4	15	14.9
2002	11.87	2.98	75.5	65	4,294	780.49
2003	7.01	2.13	52.75	35	108	1,128.12
2004	2.36	5.22	12.64	495	118,772	–
2007	15.04	6.74	108.68	134	–	–
2008	4.16	3.14	0.29	40	8,002	1,444.24
2010	–	1.47	0.25	17	3,754	–
2013	–	0.71	0.08	–	181,114	–

Source: Directorate of Economics and Statistics, Government of Assam (some of the data, especially about the total estimation of losses and data from the highly devastating floods of 2012 are not recorded in the Official Statistical Manual of the Government of Assam. The reason provided is the lack of data from the district revenue circle)

floods. The State and Central Government spend approximately INR 500–600 crore annually in disaster relief and post-disaster recovery operations. However, the real cost of repeated disasters is felt more severely at the individual and household levels. On average, 12 million people in the state live with an exposure to flood and regularly migrate to safer locations during disasters. Emerging studies have started to analyse long-term migration patterns in the region to understand how the dynamics of flood, lack of economic opportunities and globalization interplays at the subnational level (Chaliha, et al., 2012). While out-migration to seek livelihood diversification remains a complex and understudied phenomenon, it is projected that the migrations from the state follow the pattern of temporary and cyclic migrations. The majority of the temporary migration occurs during the period of flood and is responded with humanitarian assistance and relief operations regularly undertaken by the state.

The cyclic movements in the form of adaptive migration are increasingly becoming prominent in the state. A substantial number of youth from Assam migrate to other states in India as a semi-skilled labour force. The Census report of 2001 placed the net out-migration from Assam at 231,987. The cyclic migrations help individuals and families to diversify their economic opportunities and enable a reliable flow of remittances for household adaptations in Assam (Das, et al., 2009; Banerjee, et al., 2015). It is estimated that INR 5.59 billion were sent by way of remittance to individual households in Assam between 2005 and 2006 (Castaldo, et al., 2012). Most cyclic migrants migrate for 10–11 months, returning home during the festive season of harvest (*Rongali Bihu*) in the months of March and April. Interestingly, the general assembly election cycle in Assam also takes place in the month of April and returning migrants overwhelmingly participate in the election process. The average voter turnout in Assam during the 2011 Assembly election has been 76 per cent, while it was estimated to be 84.6 per cent in the 2016 Assembly elections.

Apart from these long-term migrations in the form of adaptive relocation due to riverbank erosion have also become more commonplace over the years. The 2001 census of the state provided that approximately 5.2 million people in the State of Assam have migrated at least once in their lifetime (Census of India, 2001). Since 1950, more than 4.27 lakh hectares of land has been eroded by the Brahmaputra River system, which constitutes a 7.4 per cent of the total area of the state lost to erosion in merely six decades. The annual average loss of land is nearly 8,000 hectares per year across the 17 riverine districts (Water Resource Department, 2016). In 2015, two surveys were conducted at the revenue circle level by the district administrators and by the special advisor to the chief minister to estimate the number of families living in embankments and affected by erosions. It was estimated that between 10,153 and18,495 families would need permanent relocation assistance due to bank erosions. The survey took into account only the families who are landowners and are directly affected due to their presence in erodible embankments. The survey does not account for families who are tenant residents or who have lost their lands to sedimentation instead of bank erosion (Hazarika, 2016).[1]

There is an elevated sense of acceptance within the state administration that instances of outmigration from the state are on the rise due to economic and livelihood constraints. Out of the 12 senior-level administrators interviewed for this study, 8 reciprocated strongly on the assumption of the net outflow of migrants from the state while also accepting that the 'structural development for flood protection in the State has been rather slow' (Carpenter, 2016; Hazarika, 2016). Traditionally, relief operations and post-disaster structural recovery have been two central concerns in the disaster management approach employed by the state. While the State administration is conceptually aware of the complex nexus between poverty, vulnerability and migration, the executive functioning of the state are devoid of this rationality. The senior government officials are intuitively concerned on the changing pattern of hydro-meteorological cycles due to climate change and acknowledge that the 'lack of scientific data and knowledge impedes a real conclusion and policy push on the matter'(Kumar, 2016a; Annon, 2016).

Chronic disasters and resultant displacements raise complex questions about the conceptualization, operationalization and pragmatism of state support. Conceptually, a chronic situation is problematic, as the traditional notion of moving in a linear fashion from relief through rehabilitation towards development does not always apply. The linear progression from relief to development is overshadowed by the process of planning chronic emergencies. The period of stability required for building disaster-controlling infrastructure coincides with the period of disaster and disaster relief operations. In situations of chronic disasters, a short-term thinking develops within the government which moves from planning relief from one disaster cycle to another, leaving little space for long-term infrastructure building projects.

On the operational scale, it is complicated, as areas facing chronic disasters are often marred with insecure and violent social and political landscape. Historically, the State of Assam has witnessed a high level of ethnic violence, and episodes of violence regularly erupt after the floods. The violence diverts precious resources and administrative attention from relief operations towards peace building. Pragmatically, while financial assistance to establish infrastructure in the state can enhance its resilience capacity, it is nearly impractical to imagine a state without disasters or migrations. In a poor state like Assam, which is underrepresented in a federal country, the gap between emergency and developmental programming is non-conclusive due to the restricting income generation mandates at the disposal of the State Government and the limited financial aid from the Central Government.

Flood management practices

As per Entry-17 of the List II, Schedule VII, of the Constitution of India, the flood control and management works fall within the ambit of the state legislature and the states are expected to plan and implement flood management measures through their budgetary resources. Assam adopted the Assam Drainage and Embankment Act in 1954.[2] The Act specifically provided for the construction, maintenance and management of embankments and drainage work in flood-prone regions. This Act provided specific guidelines for the operational requirements of building dams and drainage structures, and procedural requirements for acts of land requisition, contracting of works and so on. A considerable amount of planning power was vested in the district-level authorities, while the economic powers were left at the discretion of the state secretariat. The Act was replaced by the Assam Irrigation Act of 1983, which shifted the focus from embankment building towards traditional watershed management practices due to its ability to defuse manageability at the village level and provide local empowerment.[3] At the same time, the Water Resources Department at the secretariat level as the nodal agency in river management remained the primary stakeholder in implementing various structural projects for dealing with the problems of erosion and inundation in flood-prone districts. However, the state budget has remained exceedingly modest for any long-term sustainable structural enhancements in the flood plains of Brahmaputra or Barak river systems.

As there is an absence of a clear policy on the assurance of Central Government support for the infrastructural building, political parties play a vital role in negotiating the public spending and pushing their party agendas, including regional wing plans in the Central Government's funding priority. In Assam, there has been a stable political environment since 2001 under the singular leadership of Shri Tarun Gogoi, forming the Congress-led government. At the same time, the Central Government of India between 2004 and 2014 was also led by the Congress Party under the leadership of ex-Prime Minister Manmohan Singh, who was elected from the Guwahati Constituency in Assam. Theoretically, this provided sufficient space for building long-term flood mitigation projects. Factually, during the 11th Five Year Plan, the Ministry of Water Resource, Government of India, gave a financial approval to 73 schemes with a total budget of INR 432 crore. Apart from this, the North Eastern Council took up architectural enhancement plans with an estimated cost of approximately INR 20 crore and INR 250–450 crore were allocated annually towards State Disaster Relief Funds. Nevertheless, the level of infrastructural development in the region remained gradual and raised questions about the rationality of the whole process.

There is a common belief that the financial resources allocated for development are misspent due to corruption. Assam's link to corruption is as banal as it is elsewhere in India. It is somewhat recognized and accepted by the bureaucratic officers that delaying the projects is a widely used approach to gathering 'appeasement funding'. Through this method, funds from relief operations are misappropriated by the government officials and local contractors equally (2016). Every year, after the flood season, when the major construction works on dams and highways is resumed, a sit-in protest by the local population obstructing the projects is carried out. During this time, local politicians act as crowd managers and demand appeasement funding for dispersing the protestors. Given the violent history of Assam and the continued struggle for identity and integration, such protests create a hostile work environment throughout the whole state and disincentivize even the well-meaning officers and workers.

In terms of non-structural relief procedures, every year by the end of December the Government of Assam seeks tenders from the public to secure relief supplies. The contracting of relief works is *prescriptively finalized* by the middle of February (Angamuthu, 2016). There is substantial evidence that the timeline for contracting relief is somewhat non-essential; most of the districts prefer contracting by the close of March in anticipation of any early floods in the month of April. Each district contracts services for its individualized relief operation based on its past experience in flood management. Revenue circles within the state identify and prepare designated shelters. School buildings and community centre buildings are earmarked to be used as shelter by the people displaced due to floods. A standardized relief support in the form of essential food and medical supply, clothing, baby products, hygiene products, portable drinking water and much more are provided to the displaced population for a duration of three days. For the areas facing inundation for more than three days, exceptional measures are adopted to continue the flow of essential supplies. In the regions where the flood has led to decreased mobility by cutting off roadways or railways, State Disaster Relief Forces are deployed to evacuate people and livestock. In exceptional circumstances, assistance is sought from the National Disaster Relief Force. The process of providing temporary protection and evacuation assistance to the flood victims has systematically developed and internalized into the flood management policies of various state departments such as the Revenue Department, District Commissioner's Office, Public Works Department, Maternal Health and Child Welfare Department, Veterinary Health Department, Home Guards and so on (Annon, 2016). The Disaster Management Act of 2005 and subsequently arising Flood Relief Policy within the Disaster Management Manual of the Assam State Government simply reiterates the existing effective strategies of management while clarifying the precise role of different departments and officials at the state, district and local authority level. Local bodies like the Panchayats and municipalities are the first level of contact for the public during any disaster including the flood disaster, while subsequent levels of district administration, state secretariat and national government are expected to provide coordination and support on need basis.

As discussed so far, the state only has a limited policy of dealing with temporary displacements caused due to floods and requiring immediate humanitarian intervention. There is no candid policy on labour or cyclic migration. It is a widely accepted norm that the residents of Assam due to their citizenship of India have an unabridged right to freedom of movement, occupation and residence under Article 19(1) of the Constitution of India. In the philosophical and legal paradigm, the freedom of movement within one's own country has always been assumed as an undeniable part of the 'natural rights of all humans'. It is the right to travel outside one's country – immigration – which has been contentious and disputed. As a matter of executive authority, a sovereign has the right to reject admission of an alien onto its territory but a sovereign state can no longer expel its own citizen from its territory. The qualifying words in the context of the right to freedom of movement and settlement in the constitutional text are 'freely' and 'citizen', which imply the presence of an absolute right of movement available only for the citizens of India. That is, a citizen is completely free to decide where to move, how to move, when to move and for how long to move. The right ensures a guarantee of interstate as well as intrastate movement for all its citizens. An unabridged freedom to enter or move freely within the Indian Territory is not similarly available to a foreigner or a non-citizen/alien. While an alien who is legally present in Indian Territory enjoys the legal safeguard and the rights, the scope of rights available for an irregularly present alien are restricted to dignified humanly treatment, right against torture and right against deportation if such deportation would lead to persecution of the alien in question. Saving these exceptions, the Government of India has broad powers to detain and deport any person who is identified as an irregular or illegal migrant.

The right to freedom of movement is subject to restriction enacted under clause 5 of Article 19. Any sort of externment or internment order requiring a citizen to leave a certain area or refrain from entering an area or not to leave a certain area would need to fall within the qualifying test of clause 5 of Article 19, which prescribes that the restriction on the freedom must be reasonable and in the interest of general public or for the protection of the interest of Scheduled Tribes. In the *NB Khare versus State of Delhi* (AIR 1950 SC 211, 216) elaborating on the right to movement, the Supreme Court held that any externment order given by the State Government after a subjective satisfaction of the issuing executive is open to the process of judicial review except in the case of state or national emergency. The conditions in which emergency can be proclaimed by the president of the country upon the entire nation or any part of it are defined under the constitutional provisions of Articles 352–360. None of these constitutional provisions lists disasters such as those caused due to floods as a reason for proclamation of an emergency. In fact, the right to freedom of movement is a key necessity for the communities occupying vulnerable landscapes which are subjected to frequent disasters. The ability to move out of a geographical area and physically sever oneself from vulnerability is the essential element of survival.

However, migration within and outside the State of Assam has remained one of the largest controversies in the political landscape of the region. And any form of movement could be constructed with the elements of suspicion not just within the state but all across India. Most of the resources date back to the politicization of migration in Assam to the 1947 East Pakistan and India partition, during which ethnic Bengali Hindu population who crossed an imaginary boundary to become part of the newly formed Indian nation was draconized as 'infiltrators' (Hazarika, 2000; Van Schendel, 2004). Although at the time the narrative of 'infiltration' in Eastern India was dampened by the cries of holocaust in the Western Indian borders, the narrative got reignited during the Bangladesh Liberation War of 1971. Around 10 million East Pakistanis, mainly Bengali-speaking Hindus, took refuge in India during the time (Hussain, 2003). The Bengali migrants are perceived as a threat to indigenous Assamese culture on the one hand and a potential 'vote bank' on another (Weiner, 1983; Fernandes, 2005; Engineer, 1995).

Politics of migration in Assam

In 1962, after a parliamentary debate about the continued infiltration of ethnic Bengalis into Assam, attempts were made to physically seize and forcefully expel around 200,000 people across the border. This seizure and expulsion was justified under the vast discretionary powers of the Foreigners' Act at the time. However, to stop such massive push-backs and avoid confrontations with the government of Pakistan, the Home Ministry of India in consultation with its then East Pakistani counterparts, repealed the application of Foreigners Act from the State of Assam in 1964. The period from 1964 is a period of zero migration management policies in the region (Singh, 2010). Similarly, during the Bangladesh Liberation War, an unaccounted number of Bengali refugees were housed in camps across the Eastern borders. Although

The government of India maintains that the refugees returned to the newly constituted Bangladesh within one year, there is no data or record on either deportation or voluntary repatriation. The land boundary between Assam and Bangladesh remained mostly unfenced until 2012, giving a continuous ignition to the infiltration narrative and strongly influencing the landscape of governance and politics in the state.

During the 1979 by-election, objections were raised in the Mangaldoi Constituency regarding 70,000 names which were added to the revised electoral rolls. It was alleged that the names were those of Bangladeshi citizens who immigrated to India in1971. The ruling Janata Party at the time ordered an investigation, which resulted in 47,658 names in the electoral rolls to be dropped as being irregular. This led to statewide protests, and a state of emergency was declared during the 1979 Assam Agitation. The key political party in the state – United Liberation Front of Assam – alleged through popular propaganda that illegal foreign nationals from Bangladesh are migrating to Assam and changing the demographic landscape of the state, essentially turning Assam into *Bada Bengal* (Greater Bangladesh). A sense of euphoria started to emerge against the Bengali-speaking communities which have become the cause of continued segregation between Assamese community and Bengali community ever since. The period between 1980 and 1984 was a period of extreme violence in the state with statewide protests, physical damage to the infrastructure, disruption of movement and a systematic episode of violence in Nellie where 2,000 Muslims were killed in a five-hour rampage (Hazarika, 2000). At the time, the Rajiv Gandhi-led Congress Party signed the Assam Accord with a promise to cleanse the citizenship records and deport irregular Bangladeshi migrants. Bengali Muslims who were most at risk of being alleged as illegal Bangladeshi migrants took this as a major act of betrayal and withdrew their support of the Congress Party during the 1985 elections (Hazarika, 1988; Baruah, 1999; Engineer, 1995). With the result, Asom Gana Parishad (AGP) emerged as a majority party and formed a government under the leadership of Prafulla Kumar Mahanta. However, the non-cooperation of the Congress Central Government led to AGP's failure in its efforts to identify and repatriate Bangladeshi migrants, causing internal dissatisfaction and fraction within the party.

The Congress Party at the centre to appease the Assam movement leaders adopted the Illegal Migrant (Detention and Termination) Act, the constitutionality of which was successfully challenged in the Supreme Court of India. The act nevertheless allowed for identification of illegal migrants by shifting the onus of proof of citizenship on the person who has been alleged to be a non-citizen. While this act brought the Congress Party back into power, it did little to suppress the anti-migrant sentiments in the state. The sentiments eventually got diversified and created larger rifts between several ethnic and religious sub-groups. In 1994, native Bodo Tribes clashed with the ethnic Bengalis in Barpeta area affecting 3,500 migrant families and forcing 1,500 people into relief camps. The Indian Army was called in to handle the situation, and the failure of the Congress Government to respond effectively was severely criticized. Once again in a placation strategy, the Bodo Autonomous Tribal Council was created and was given rights to govern the historical Bodoland. Unsurprisingly, Kokrajhar which serves as headquarters for the Council is a Muslim-majority area. It is worth noting that the year preceding 1979 which saw the birth of Assam Agitation there were two highly devastating floods in 1977 and 1978 displacing close to 100 million people. Although the effect of these floods on the agitation typology is not studied, it is an area worth pondering.

A multitude of ethnic and religious factions currently occupying the political landscape of Assam have made it highly possible for any person engaged in intrastate migration as being lamented as an 'illegal migrant'. The ability to prove one's citizenship in a nation which until 2014 had not made registration of births and deaths a mandatory exercise comes with heavy social and economic cost. This point is elaborated through the field surveys and group discussions conducted in *chors* and riverbanks in the districts of Morigaon, Dhubri and Goalpara. Bhashayani *Chor*, a 30-minute boat ride from the Dhubri riverside, is a region thathas faced continuous disappearance of landmass including the full submergence of a river island in the 1978 floods. People inhabiting this island took refuge in the Dhubri riverside region and nearby *chors* coming under the South Salmara revenue circle. Unable to return to their original place of residence, these people built their lives afresh in their newly habituated settlement without any specific assistance from the government. As the Dhubri district has a multiplicity of closely lying *chors*, the living conditions within these *chors* are very similar there is to an existence of social capital. The *chor* inhabitants allow each other to occupy a part of their lands through shared land tenancy, through which subsistence yields can be produced for use by the displaced families. Similarly, in the *eroded area* in Laharighat revenue circle of Morigaon, families have moved to the interior villages of Laharighat revenue circle (falling within Morigaon district) and Dhing revenue circle (falling under the Nagaon district). Those who moved to Laharighat revenue circle continue to remain under the same administrative jurisdiction, while those who migrated to either Nagaon district in the East or to Darrang district in the North might have to re-register as new voters in the new areas. As the Bangladeshi immigration issue is at the heart of Assamese politics, its civic life and governance, it is highly likely that in the absence of a reliable social capital these flood victims would be seen as 'immigrants'. And the ethnic and religious composition of the displaced population would determine their fate in their new host society. The process of identification and registration as a voter would require the production of legacy data in the form of a government-issued certificate of birth, death, marriage and so on. Inability of providing these documentations would lead to being classified as a doubtful citizen or a D-voter.

It has been seen in the Asudubi and Dhobakura villages of Goalpara district that several women who migrated due to loss of their lands to floods were pushed to the D-voter list in their new constituency. Although they were able to arrange their legacy data eventually, it would take them on average 10 years and INR 8,000 in court proceedings to get their status corrected by the Foreigners' tribunal. A lot of the permanent relocation in the state occurs due to complete submergence of river banks and river islands and all those migrating due to such submergence risk losing their rights as a citizen of India. At present, there are 137,601 D-voters in Assam, 68.7 per cent of them being women. Table 5.2 provides a statewide distribution of D-voters in Assam. While the creation of the D-voter list which shifts the burden of proof on the person alleged to be an illegal migrant itself is ultra vires the doctrine of due process of law, what is even more startling is the fact that the entire families of D-voters are ostracized from the government-run welfare programmes. The field officer in Dhobakura, who refused to provide his name and is supervising the updation of the National Citizenship Registry in the Jaleswar constituency circle, commented that 'D voters are as good as a stateless person. They cannot claim any benefit from welfare or rehabilitation schemes. They can't even seek PRC for higher education'. It is worth mentioning here that the right to primary education and the right to seek humanitarian assistance are basic human rights available to all people within a country irrespective of their citizenship status. Similarly, right against torture or undignified treatment is also available to the D-voters. However, rights to seek political representation, right to assemble or peacefully protest and right to overcome their vulnerability by migration or through seeking rehabilitation assistance are some of the rights unavailable to anyone listed as a D-voter.

Table 5.2 Distribution of D-voters across Assam

District	Total number of D-voters			Removed
	Total	Male	Female	
Dhubri	11,875	3,210	8,665	185
Goalpara	10,389	3,056	7,333	481
Morigaon	5,433	2,466	2,967	68
Kamrup M	4,743	2,250	2,493	80
Dhemaji	4,367	2,093	2,274	30
Bongaigaon	1,247	674	573	22
Chirang	233	75	158	0
Jorhat	17	11	6	0
Sonitpur	24,538	9,835	14,703	55
Barpeta	22,477	7,543	14,934	517
Nagaon	15,255	5,444	9,811	54
Udalguri	10,169	4,732	5,437	236
Cachar	5,847	2,672	3,175	190
Darrang	5,324	1,872	3,452	69
Baska	2,757	1,068	1,689	0
Karimganj	2,619	728	1,891	0
Tinsukhia	2,512	1,245	1,267	25
Goalaghat	2,429	1,131	1,298	0
Lakhimpur	2,304	979	1,325	11
Kamrup Rural	1,596	443	1,153	27
Kokrajhar	1,073	467	606	15
Karbi Anglong	1,041	458	583	46
Dibrugarh	774	418	356	7
Nalbari	631	325	306	7
Hailakandi	102	90	12	31
Dima Haso	5	4	1	0
Total	139,757	53,289	86,468	2,156

Source: Authors' own calculations based on the revised D-voters list, 2016. Election Commission of Assam Data, received on 04.04.2016, Guwahati D. C. Office

Politics of rehabilitation in Assam

There is a policy in the state for granting ex gratia compensation to families who have suffered either the loss of agricultural land or household or both under the State Government Flood Relief Scheme. However, to seek compensation for loss of land, the landownership record must be submitted by those claiming damages. Maintaining property rights in a state like Assam, where the land holdings are small and for the purpose of subsistence farming alone, are neither easy nor economical. The Brahmaputra River has a multiplicity of river islands, and every year the river braids during the monsoon season and creates new *chors* while destroying a few old ones. The eroded *chors* get de-habituated while the new *chors* get inhabited by a somewhat floating population. To maintain land rights, the landowners have to continue paying taxes on the piece of land recorded under their names in the cadastral data. It is entirely possible that the property in question may no longer exist or is no longer habitable due to sedimentation. However, the payment of land revenue must not stop because that will vacate the ownership rights and any future compensation that the owners might be entitled to would also be divested.

Multiple claims may arise on a piece of land which may re-emerge during subsequent floods if such re-emergence goes unnoticed by the original *patta* holder. Oswin Nampuri, circle officer from Kamrup Metro, explains that such claims are regularly brought from the *chor* areas as both the land and population are floating. Narrating one of the recent cases, he said that once a land has eroded the *Lot Mandal* makes an erosion report, based on which erosion victims can claim compensation or rehabilitation from the state revenue department. These claims along with the certificate of ownership are attested at the revenue circle level and approved by the district commissioner. An ex gratia payment ranging from INR 50,000 to INR 200,000 is made based on the type of land in question. And rehabilitation land for construction of homestead is approved if the district is able to identify a suitable wasteland for the purpose. However, it is possible that long after such compensations are made, the land may re-emerge and the original owner may try to assert rights. Meanwhile, it is also possible that by virtue of adverse possession, some new resident might already have occupied the land in question. Under the land laws for the state, a property can be occupied through adverse possession after 12 years of continued and uncontested occupation. In case of a government land, adverse possession requires occupation of 30 years. To prove occupation, one needs to keep paying taxes on the same piece of land, and as mentioned above it is possible that such land may no longer exist due to flood erosion; but as substantial monetary rights are tied to the land ownership, maintaining the same by regularly paying land revenue remains important.

Although most of the officials interviewed empathize with the disaster victims, they also reflect a myriad of mixed opinions. For example, Shri Rajendra Kumar, who retired as the commandant general of the

Home Guards and director of the Civil Defence in the State of Assam, explained that the 'disaster victims are often self-styled victims' who would seek 'floods as an opportunity to seek monetary compensation packages'. He explains that the 'government is bound to provide ex gratia financial support to anyone who has suffered flood damage to life or property. Often people choose to live in flood-prone areas and engage in a calculated risk in order to obtain the ex gratia support. Part of these payments is then utilized to buy cheap land in some other vulnerable region, and the whole cycle of compensation repeats in loop' (Kumar, 2016b). Several other officials agreed with the prevalence of this phenomenon of using ex gratia payment as a 'money making opportunity' (Angamuthu, 2016; Chowdhary, 2016). However, the majority of them reflected a humanitarian stance in the midst of rampant poverty and laagered development, as is evident from the remark made by another bureaucrat, 'Yes it is possible some clever people do this, but it is not right to call these people opportunists in the real sense. They are living in chronic poverty, and this is just one of the means by which they can overcome some of their hardships' (Islam, 2016).

Similarly, benefits under welfare schemes such as the Mahatma Gandhi National Rural Employment Guarantee, Indira Awas Yojana and more are tied to the citizenship status. A person's inability to produce documents in a district where such person has newly arrived will not only put them in the D-voters list, but will also restrict them from benefiting from the state-administered programmes. This essentially creates a situation of being tied to a vulnerable landscape because moving out of it might be more disastrous for a poor family dependent upon state assistance.

It is worth contemplating that in a State which regularly deals with the problem of chronic poverty why an absurd amount of time and money is invested in creating administrative systems that are nonfunctional, irregular and farcical? The State of Assam had a stable leadership under the Congress Party and a relatively uncomplicated dynamics with the Central Government which was also governed by the United Progressive Alliance with the Congress Party at its core from 2004 to 2016; yet, the level of development and efficacy in flood response has mirrored itself over the years.

Conclusion: making sense of the irrationality

The question arises: how do we decipher the situation of Assam? There is a façade of legal frameworks and soft-law instruments based on the principles of welfare, justice and equity, and yet the State Government has remained inept in preventing or managing disaster and disaster-induced migrations. Why is it that a supposedly democratic welfare state fails to provide equal protection to its communities from even the known chronic disasters?

To explain this dilemma, it is necessary to have a closer look at how institutions work in reality. Most policy analyses develop from the assumption that the rational choice actors–elected policymakers-will transform parts of their agendas into policy by creating coherent and rank-ordered set of preferences which will be delivered by the neutral executive institutions. In this starting point, we tend to either completely ignore or marginalize the organizational character of the executive agency. Executive agents seldom utilize ranking preference and they often trade off their willingness to help a victim with their aim of delivering status quo decisions. The aim is not only to provide assistance to those displaced due to floods but also to ensure that a general situation of law and order is maintained. After decades of violence and political instability in Assam, the current status quo is desirable from a bureaucratic point of view.

The concept of incrementalism in public policy making recognizes that an actor may take many small steps of change and few steps of hyper-incrementalism. It is normatively desirable for the organizations to work through incremental behaviour instead of adopting rational choices. Public policy, when conducted in a progression of small alterations, allowed lesser risk and levelled progression towards the major policy objective. Policymakers working towards their aim, in an incremental approach, can better analyse unforeseen externalities by understanding exactly which alteration to the plan produced them. In the long run, this can lead to policies that better address the particular issue in question without imposing 'serious lasting mistakes'. The incrementalism in the executive functioning accounts for the unpredictability in the political system. By creating an elongated path dependency in its operation, the system tends to avoid and overcome violent outbursts fuelled by the political narratives surrounding migration.

This is visible through the functioning of the state apparatus in Assam. The divisional politics in the state continues to flourish on the narrative of illegal migrants, and the political factors continue to stigmatize the migrant community. Political dynamics of securitization in the case of migrations tend to identify the category of people who should be viewed as an 'outsider', 'non-citizen', 'illegal migrant', 'foreigner' or 'other' and must be excluded from welfare benefits. The executive wing, on the other hand, has taken several small steps to ensure development of Flood Early Warning System and integration of relief response by standardizing the type and duration of relief process. Similarly, acting towards job creation with the support of the Central Government under the National Rural Employment Guarantee Program instead of adopting a migration management policy at executive level showcases incrementalism and integrationist approach adopted by the system.

In divided societies facing internal migration issues, local governments tend to ignore the murky business of migration management, relocation and resettlement, and concentrate towards the linear trajectory of relief to development, even if it is pragmatically unachievable. The responsibility of relocation and resettlement is shifted to the victims of disaster who are proclaimed as dynamic actors having the freedom to migrate and reside anywhere within the national boundaries (Gogoi, 2016; Annon, 2016). Even in scenarios where a new policy guideline may require response towards risk reduction and planned relocation, the administrative set-up may only act in an incremental manner and not change the status quo drastically. And although there are ethical, moral as well as accountability questions towards disaster management and displacements in the state, the state executive is increasingly aware of its role in managing the same through inter-agency and at times interstate transactions.

Acknowledgements

The author is thankful to the State Government of Assam, especially to the ex-Chief Minister Shri Tarun Gogoi for providing insights and data for this research. Interpretations and opinions reflected in this work are sole responsibility of the author.

Notes

1 Report on 'Landless Families Living in Embankment' submitted to the chief minister of Assam pursuant to secretarial communication dated 28 January 2016. Copy of the communication and the report provided by Assam State Disaster Management Authority.
2 Assam Act No. 1 of 1954. The text of the Act can be downloaded from http://faolex.fao.org/docs/pdf/ind4229.pdf
3 Assam Act No. VIII of 1989. The text can be downloaded from http://faolex.fao.org/docs/pdf/ind90361.pdf

References

Angamuthu, M., 2016. Deputy Commissioner, Kamrup M [Interview] (17 February 2016).

Annon, I., 2016. Interview no. 1 [Interview] (15 February 2016).

Banerjee, S., Bisht, S. & Mahapatra, B., 2015. Building adaptive capacity in Assam. *Forced Migration Review*, 49, pp. 66–68.

Baruah, S., 1999. Ethnic conflict as stat – society struggle: The poetics and politics of Assamese micro-nationalism. *Modern Asian Studies*, 28(3), pp. 649–671.

Carpenter, V., 2016. Interview no. 10 [Interview] (23 February 2016).

Castaldo, A., Deshingkar, P. & McKay, A., 2012. *Internal Migration, Remittances and Poverty: Evidence from Ghana and India*. Falmer: University of Sussex.

Census of India, 2001. *Migration Tables*. Registrar General and Census Commissioner, Ministry of Home Affairs, Government of India, New Delhi.

Chaliha, S., Sengupta, A., Sharma, N. & Ravindranath, N. H., 2012. Climate variability and farmer's vulnerability in a flood-prone district of Assam. *International Journal of Climate Change Strategies and Management*, 4(2), pp. 179–200.

Chowdhary, K., 2016. Project Officer, District Disaster Management Office, Chirrang [Interview] (29 February 2016).

Das, P., Chutiya, D. & Hazarika, N., 2009. *Adjusting to Floods on the Brahmaputra Plains*. Assam, India, Guwahati: ICIMOD/Aryanak.

EM-DAT, 2016. *The CRED/OFDA International Disaster Database*, Brussles: Université Catholique de Louvain.

Engineer, I., 1995. Politics of Muslim vote bank. *Economic and Political Weekly*, 30(4), pp. 197–200.

Engle, L. Nathan, Owen, R. Johns, Lemos, Maria Carmen and Nelson, Donald R., 2011. Integrated and adaptive management of water resources: Tensions, legacies and the next best thing. *Ecology and Society*, 16(1), pp. 1–11.

Fernandes, W., 2005. IMDT Act and immigration in North-Eastern India. *Economic and Political Weekly*, 40(30), pp. 3237–3240.

Fussel, H. M., 2007. Adaptation planning for climate change: Concepts, assessment approach and key lessons. *Sustainability Science*, 2(2), pp. 265–275.

Gogoi, T., 2016. Interview no. 26 [Interview] (16 April 2016).

Goswami, B. N., 1998. Interannual variations of Indian summer monsoon in a GCM: External conditions versus internal feedbacks. *Journal of Climate*, 11(4), pp. 501–522.

Hazarika, N., 1988. Asom Gana Parishad. *The Indian Journal of Political Science*, 49(1), pp. 95–104.

Hazarika, N., 2016. Interview no. 3 [Interview] (14 February 2016).

Hazarika, S., 2000. *Rites of Passage, Border Crossing, Imagined Homelands, India's East and Bangladesh*. New Delhi: Penguin Books.

Hussain, W., 2003. Bangladeshi migrants in India: Towards a practical solution. In: P. Chari, M. Joseph & S. Chandran, eds. *Missing Boundaries*. New Delhi: Institute of Peace and Conflict Studies, pp. 125–150.

Islam, N., 2016. Deputy Commissioner, Dhubri [Interview] (1 March 2016).

Kotoky, P., Bezbaruah, D., Baruah, J. & Sarma, J., 2003. Erosion activity on Majuli: The largest river island of the world. *Current Science*, 84(7), pp. 929–932.

Kumar, R., 2016a. Interview no. 7 – Deputy Commissioner Morigaon [Interview] (17 February 2016).

Kumar, R., 2016b. Retired Additional DIG, Assam Police [Interview] (23 January 2016).

Mandal, S., Choudhury, B. U. and Satpati, L. N., 2015. Moonsoon variability, crop water requirements and crop planning for kharif rice in Sagar Island, India. *International Journal of Biometeorology*, 50(12), pp. 1891–1903.

Ravindranath, N. H. et al., 2011. Climate change vulnerability profiles for North East India. *Current Science*, 101(3), pp. 384–394.

Shukla, S., 1997. *Transforming the Northeast (High Level Commission Report to the Prime Minister)*. New Delhi: Government of India.

Singh, D. K., 2010. *Stateless in South Asia*. New Delhi: SAGE.

Singh, S. K., 2007. Erosion and Weathering in the Brahmaputra River System. Large Rivers: Geomorphology and Management, pp. 373–393.

Van Schendel, W., 2004. *The Bengal Borderland: Beyond State and Nation in South Asia*. London: Anthem Press.

Water Resource Department, 2016. Flood and Erosion Problem. [Online] Available at: http://assam.gov.in/web/department-of-water-resource/flood-and-erosion-problem

Watmough, G. R., Atkinson, P. M., Saikia, A. & Hutton, C. W., 2016. Understanding the evidence base for poverty – environment relationships using remotely sensed satellite data: An example from Assam, India. *World Development*, 78(2), pp. 188–203.

Weiner, M., 1983. The political demography of Assam's anti-immigrant movement. *Population and Development Review*, 9(2), pp. 279–292.

6 Remittances as self-insured life

On migration, flood and conflict in North-Western Pakistan[1]

Giovanna Gioli

Remittances, resilience and self-insured life

Researchers are offering new evidence and frameworks to investigate the nature of the relation between human security and the environment. Since the end of the Cold War, we have witnessed a 'broadening of the concept of security from its traditional focus on states to include the many *economic, political, social and environmental factors that menace the sustainability of life and livelihoods*' (UNDP 1994, cited in Duffield 2010: 63; emphasis added). Under the policy imperative of enhancing vulnerable populations' resilience to a variety of shocks and stressors, idioms such as 'climate resilient peacebuilding' (IISD 2010)[2] or 'climate and conflict resilience' are gaining ground in today's global governance. Several national and international agencies are directing their efforts towards 'consolidating existing knowledge regarding climate change, fragility and conflict and examine emerging interlinkages in the field' (Tanzler et al. 2013) in order to develop 'conflict sensitive responses to climate change' (Vivekananda 2011).

The contours of the pervasive concept of resilience to global environmental change are becoming more and more encompassing of other kind of shocks and disturbances. Resilience hence denotes 'the ability of countries, communities and households to anticipate, adapt to and/or recover from the effects of potentially hazardous occurrences (natural disasters, economic instability and conflict) in a manner that protects livelihoods, accelerates and sustains recovery, and supports economic and social development' (Food and Agriculture Organization [FAO] 2012). As vulnerability to crisis is intrinsically shaped by social, economic and political factors, as articulated by global inequalities and uneven access to and control over assets and resources (Wisner and Luce 1993), most of the times large-scale natural disasters and conflict occur in the global South, where informal work, precarious land rights, subsistence agriculture along with the lack of access to financial instruments and products and other forms of social protection severely limit the ability of people to cope with crisis and insure themselves against risks.

If we look at the subset of literature addressing the nexus between migration and adaptation to climate change, we can detect a notable and comparable 'broadening' of the issue. The framing of the nexus between migration and climate change has shifted from an open 'securitization' (Suhrke 1994; Myers 2002; Stern 2007) and a narrow focus on environmental pull factors and forced migration to what can be called as its 'developmentalisation' (Felli 2013; Bettini 2014; Bettini and Gioli 2016). This involves the idea that *labour* migration can represent a legitimate and positive adaptation strategy to (global) environmental change (Foresight 2011; Tacoli and Mabala, 2010; Black et al. 2011). The 'migration as adaptation' thesis can be considered a subset of the 'migration and development' literature that is conceptually grounded on the New Economics of Labour Migration (NELM). NELM pivots around the role of social and financial remittances and represents migration as a grand bargain among sending and receiving areas and migrants themselves (Stark and Levhari 1982). Households rather than an individual become the unit of decision-making (Stark and Bloom 1985: 175). The implicit contract among these actors is that migrants are bound to remit. Spatial diversification of labour and resource pooling help the household to mitigate risks (Arango 2000) such as crop failure caused by changing weather patterns or environmental shocks. Migration as a risk-minimizing strategy acts as a form of insurance or social welfare for those countries where private insurance mechanisms are unavailable, imperfect or inaccessible to poor families (Massey et al. 1998). Thereby, remitted assets cushion risk and might have a transformative impact, especially in rural economies (Taylor and Martin 2001).

Drawing on Duffield's biopolitical dichotomy between 'developed' and 'underdeveloped life', we contrast the global North depending upon welfare bureaucracies and infrastructures (*developed*, insured life) with the global South where the informal life supports derived from family and community are a substitute for social protection (Duffield 2010). With the permanent crisis of 'adaptive self-reliance' (Duffield 2010), and in line with NELM, households shield themselves from increasing risk by diversifying their portfolio of livelihoods and resorting to labour migration as a form of 'household sponsored insurance system' (Yang and Choi 2007) and a substitute for social security (Schrieder and Knerr 2000).

In time of economic and environmental shocks as well as during and after conflict, remittances constitute a form of 'self-help' from below. As recently put by IOM: 'In the context of natural or man-made catastrophes and crises, remittances and migration can support the resilience of populations both staying and going' (IOM 2014). The rationale behind this enthusiasm is the surging amount of global remittances, which reached about USD 414 billion in 2013 (World Bank 2014), more than three times the size of the ODA. Moreover, they tend to act as counter-cyclical shock absorber. While capital flows and other types of foreign exchange tend to rise during favourable economic times and fall during less favourable ones, remittances are stable and often act counter-cyclically to other forms of investments (Ratha 2003; Kapur 2005; Higgins et al. 2004; Dorantes and Pozo 2006; World Bank 2006).[3]

Evidence from conflict-affected areas (e.g. Bosnia and Herzegovina, Kosovo, Sri Lanka, Afghanistan, Somalia, Liberia, Côte d'Ivoire) show that migrants transfer funds and invest in their countries of origin when international investment has disappeared (Sriskandarajah 2002; Bisogno and Chong 2002; Fagen and Bump 2006). In many contexts, remittances proved to be a necessary condition for peace and rebuilding. Similarly, remittances were found to increase strongly in response to adverse exogenous shocks such as natural disasters or large declines in the terms of trade (Mohapatra et al. 2009). After the 2005 earthquake in Pakistan, households with international migrants were found to be better able deal with emergency needs and recovered faster and better such as by more quickly rebuilding their house, restarting agricultural production and paying school fees than the community members with no or reduced access to this resource (Suleri and Savage 2006). In the Samoa, remittances increased significantly after the 2012 cyclone and the 2009 tsunami (Paulson 2003; Le De et al. 2013), and they proved crucial for recovery after the 2004 Asian tsunami (Laczko and Collett 2005).

The contribution of this chapter to the literature debate on 'remittances in times of crisis' is twofold. On one hand, the case study discussed here aims at shedding some light on the perceived role of remittances among the conflict-affected and disaster-prone communities of the Swat and Lower Dir Districts of Khyber Pakhtunkhwa (KP). Local perceptions on the importance of remittances during and after the Taliban conflict (2009) and the 2010 floods are compared and analysed against the framework of NELM. On the other hand, the results are critically assessed with reference to the literature on 'migration and development' and 'migration and adaptation'. Section 2 provides an overview of labour migration in Pakistan and in the study area and a description of the research methods. Section 3 describes the overall characteristics of migration and the perceived importance of remittances at both household and community level. In section 5, the role of remittances during the 2009 conflict and 2009 floods are presented, while section 6 critically discusses the results against the background of the debate on migration and development. In section 7, we present our conclusions and policy recommendations.

1. Human mobility between conflict and (global) environmental change in North-Western Pakistan

Pakistan has the world's seventh largest diaspora and ranked seventh in 2012 for remittances inflow received by its emigrants (World Bank 2012). The inflow of remittances exceed USD 13 billion in Fiscal Year 2012 (State Bank of Pakistan 2012), featuring a 10-fold increase since 2001 (Amjad et al. 2012). Remittances from the Gulf dwarfs the total aid as percentage of the GDP (see Fig. 6.1).

According to the estimates of the Bureau of Emigration, about four million Pakistani migrant workers live around the globe (2.5% of the total population). Out of this, about half of the labour migrants work in the six countries of the Gulf Cooperation Council (Government

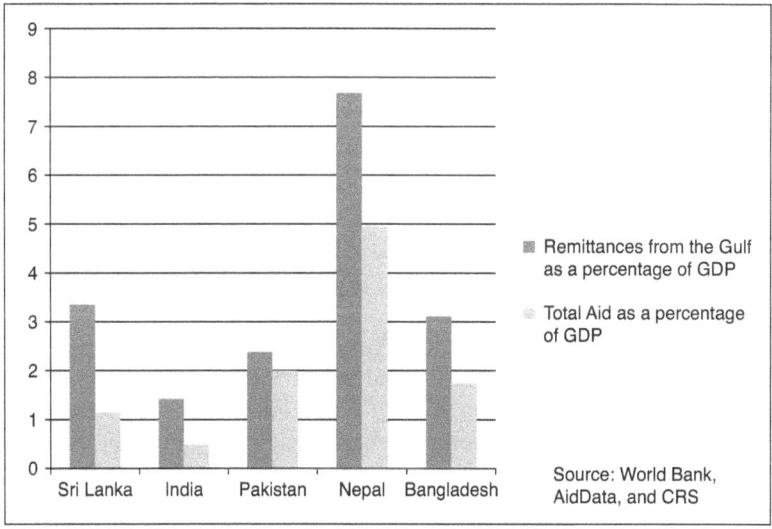

Figure 6.1 Remittances from the Gulf countries
Source: World Bank (2014)

of Pakistan 2013), primarily in Saudi Arabia and in the United Arab Emirates (UAE). Since the first oil crises in 1973, rural regions of low agricultural productivity such as the rain-fed areas of KP are the main sources of unskilled and semi-skilled migrant workers to the Gulf (Addleton 1984; Gazdar 2003; Nichols 2008; Siegmann 2010).

Migration from KP to the Gulf follows pre-existing networks with construction companies that in the 1970s and 1980s allowed Pashtuns – the major ethnic group in the KP province – to be recruited to the Middle East at a rate two to three times their population percentage within Pakistan (Nichols 2008: 145). Political factors played a role too, as after the fragmentation of Pakistan and the subsequent separation of East Pakistan as independent Bangladesh, the Pakistani government and Prime Minister Bhutto decided to embrace Middle Eastern economic and cultural connections (Nichols 2008) and *Dubai chalo* ('going to Dubai') became the most resorted answer to cope with insufficient livelihood options and to cushion the economic shocks coming from the uncertain and risk-prone agro-pastoral economy of the region (Ahmed 1981). Recurrent famines were common until a few decades ago, especially in the months between *Kharif* and *Rabi* (April–May), when food stocks are frequently over, yet it is too early to harvest. Since the 1970s, migration to the Gulf has become one of the most resorted and successful livelihood strategies, and today KP is the most dependent area on foreign remittances within the entire country: foreign remittances account for 9.4 per cent of the average monthly household income, compared to 5.1 per cent for Punjab, 1.5 per cent for Baluchistan and 0.7 per cent for Sindh (Kock and Sun 2011). KP is also the biggest recipient of remittances from abroad as well as from within the country among all the four provinces of Pakistan, and nearly 31 per cent of all households received remittances in 2007–2008, with this proportion up to 36 per cent for the rural sector of the province (Khan and Khalid 2011). As of 2010, less than 20 per cent of the total land of the province is cultivable and less than 15 per cent is currently sown (*Ibidem*).

International labour migration from KP has specific characteristics that are rooted in the traditional organization of its society. The organization of the Pashtun society relies on complex interconnections of family bonds and honour where duties and belonging are felt very strongly (Barth 1959; Lindholm 1996). As described by Watkins (2003), Pashtun migrant workers in the Gulf practise a 'save there, eat here' strategy. The common saying *alta gatee, dalta ookhree* put in a spatial/temporal dichotomy the act of saving carried out by the individual migrants *there* (in the Gulf) and the act of consumption, the 'eating' of remittances as performed by the household (*here*), which transform them into 'things of value, namely families and relationships, and the importance of these kin relations to households' (Watkins 2003: 61).

In adherence to the code of *purdah* (gender segregation) observed in the region, the totality of the labour migrants employed in the Gulf are males (Arif 2009; Siegmann 2010). Migration is prevalently temporary and cyclical, with periods spanning over decades. Migrants keep their ties with the community with the perspective of improving their household, investing there and eventually going back. Migration is so embedded in the socio-economic structure of these communities that it has acquired a role in the social hierarchy, bringing status to migrant-sending households. Fathers migrate and wait for their sons to reach the adult age in order to perpetuate migration (see Hagen-Zanker et al. 2014). The suffering and the sacrifices of migration to the Gulf are even vividly portrayed in the rich oral Pashtun culture, through both music and poetry.

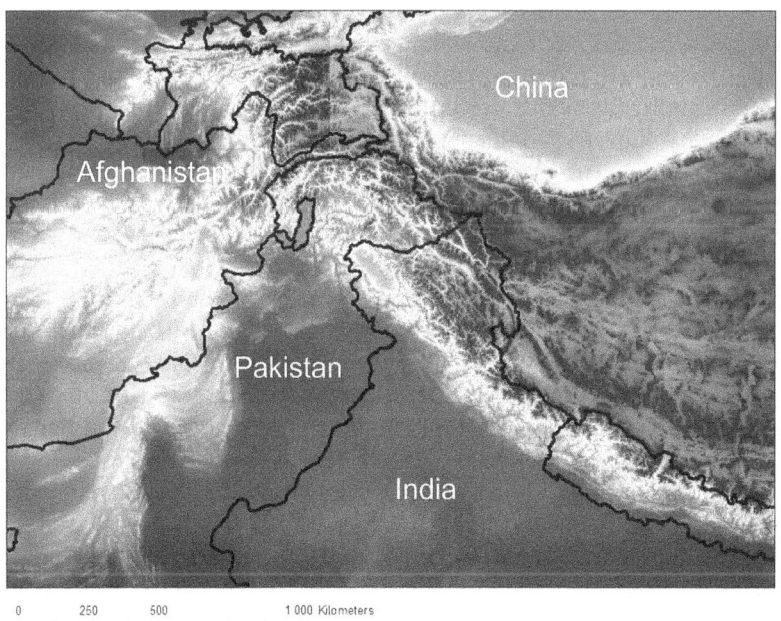

Map 6.1 Study areas (Swat and Lower Dir districts)

Source: Map drawn by Faisal Mueen Qamer

Migration to the Gulf is a steady and increasing phenomenon that has significantly improved the food security and the overall life standards of migrants sending remittances to households in KP, the second most food-insecure province in Pakistan after Baluchistan.

This study is based on a combination of qualitative and quantitative research carried out in October 2012 in conflict-affected villages of Swat and Lower Dir districts of KP, known to have high numbers of households with incidence of labour migration out of their native districts within or outside the country. Swat and Lower Dir feature some of the highest figures of workers *registered* for overseas employment (17,995 and 19,196, respectively) in 2012 (GoP 2013). Lower Dir has a population of about 1,150,000 (urban population 6%) and Swat of about 2 million (urban population 26%). Livelihoods are characterized by high prevalence of family farming, widespread dependence on natural resources and high sensitivity to environmental and climatic changes. These fragile socio-ecological systems lie in the Hindu Kush mountain range, face heavy environmental pressure and are particularly subjected to climate change adverse impacts in terms of both large climatic variation (IPCC 2007) and expected impacts on food security and livelihoods (Ericksen et al. 2011). They face harsh weather conditions such as extensive snowfall in winter and drought in summers and are particularly prone to natural hazards because of a high physical vulnerability (weak infrastructures, deforestation) as well as social vulnerability (poverty, conflict-stricken areas with archaic power structures featuring entrenched inequality).

The Union Councils were identified through purposive methods via consultative meetings with local NGOs and civil society representatives. A structured questionnaire was administered to 602 remittances-receiving households in four Union Councils, one in each of the two selected districts (Bar Aba Khel in the Kabal *tehsil*;[4] Baidara in the Matta *tehsil* of District Swat; Haya Serai and Lal Qilla in the *tehsil* Balambat of the Lower Dir district). The sample was stratified by gender in order to give some representation to women (male and female representation is 69% and 31%, respectively). The questionnaire aimed at collecting information on the perceived role of remittances during the peak of the Taliban insurgency (2009) and the devastating 2010 flood. In our sample, 22 per cent of the surveyed households have been directly affected by the 2010 flood, 90 per cent by the conflict, while 9 per cent were not affected directly by either. Two focus group discussions (FGDs), one in each district, with male stakeholders from the community were also conducted along with in-depth interviews with community leaders and migrant recruiting agents.

2.   Results and discussion

Migrant profile

The majority of the migrants were employed in the Gulf (92%), with Saudi Arabia being the top destination. Four per cent migrated to Western countries, 2.5 per cent migrated internally, while only 1.5 per cent migrated to Far East countries (Malaysia). This is due to several reasons, including the fact that Saudi Arabia has a long history of friendship with Pakistan and that obtaining a visa for the Gulf is easier and less expensive as compared to Western countries. Due to the prevailing patriarchal norms and rules, exclusively men migrate for labour; and such situation was mirrored in our sample, where the totality of the migrant workers were males, migrating alone and leaving the extended family behind (95%).

The migrant educational status was found to range between illiterate (22%) and 10th class (59%). As for the type of employment, almost half of the migrants (43%) engaged in unskilled labour, 25 per cent in semi-skilled labour, 25 per cent are skilled labourers, while only 4 per cent and 3 per cent are agricultural labourers and professionals, respectively.

Enabling social networks prove to be crucial in facilitating the migratory process, as 59 per cent of the respondents were able to reach the destination country thanks to visa facilitation provided by a relative already in the destination. Thirty-seven per cent recurred to a recruiter, while the remaining reported to be without a permit.

Remittances

Hundi/Hawala[5] is the most common way of sending remittances back home (59%), followed by bank (26%) and compatriots working in the destination (8%). Telebanking/*easy paisa* is available only for internal migration, and 5 per cent of the respondents were found to not be able to remit as they are still repaying the costs of migration.

Remittances are perceived as having played a major role in improving the living standard and the lifestyle of migrant-sending households. Remittances are predominantly used to meet the basic household needs, yet they also fund investments. In harmony with previous research in North-Western Pakistan (Steimann 2005; Siddiqui and Mahmood 2005; Suleri and Savage 2006), the first investment causing an immediate change of lifestyle is the construction of a new house. This means switching from the traditional *kacha* structure (walls made from a mixture of mud or clay and straw) to *pakka* houses (brick or concrete walls). As the construction material is quite expensive, *pakka* houses confer also social status to the household, improve the general health conditions of the family as well as the resilience to natural hazards (e.g. to floods).

Interviewees were well aware of the point to which the local economy relies on remittances: they reported that the construction industry, businesses and the transport and banking sectors depend to a great extent on remittances inflow, and they showed rising concern over a possible future drying out of remittances due to the decrease in demand for unskilled and semi-skilled labourers in the Gulf. The loss of confidence on the local resources was also mentioned as a result of their over-reliance on foreign money.

Thanks to migration, many households managed to get an electric water pump and other labour-saving technologies, thus alleviating the workload for women (see also Siegmann 2010). Remittances-receiving households could also buy modern appliances (e.g. refrigerators, washing machines, television), either brought from abroad or purchased locally. Such appliances are reported to have increased the household well-being and exposure to the outside world by both men and women interviewees. In the last five years, there is also an increasing trend in the installation of uninterruptible power supplies and generators.

Better education for children is another major change resulting from migration. As put by one interviewee in Swat:

> Thanks to remittances I am studying at university and my young brothers and sisters go to private schools. In the near past you could hardly find students from our area in colleges and university but today you will find them, all because of international remittances.

The role of migration in boosting education is two-fold: on the one hand, parents can afford to send children to school thanks to the increased income coming from remittances; and on the other, one of the main reasons why education is felt as a priority by parents is that an educated person can get a good job abroad. Remittances may improve human capital, but when such improvement is not matched by local economic growth, the result is a vicious circle of dependence upon further outmigration (Skeldon 2007). This is why education is often promoted having in mind the future of male children as better off migrant workers, as migration tends to be perpetuated from generation to generation:

> We want our (male) children to get an education so that they can get better job in the Gulf.
>
> *(Male participant, FGD in Swat)*

Migration has partially improved education for girls too, and recent estimates report an increase of female students in Swat going from 74,900 in 2009 up to 118,500 in 2011 (Ali 2013). The increased demand for education has also triggered the number of private schools: In the 1980s, 3 private schools were active in Swat, whereas today there are more than 500,[6] operating even in small villages, offering co-education (boys and girls) up to the primary level. Education of girls is also triggered by local resistance to the fact that the Taliban have been specifically opposing and targeting female education. Still, restrictions on girls' mobility, the lack of transport facility and of separate toilet infrastructures in public school along with shortages of qualified female teaching personnel are issues still faced by many villages in both Swat and Dir (Siegmann and Sadaf 2006). Additionally, damages to infrastructures and schools' facilities caused by the 2010 flood and other environmental hazards, as well as by the Taliban insurgency, have added additional hurdles to mobility and access to education.

The possibility of investing more money in social spending is also perceived as a major asset by remittances-receiving households. Such response has to be understood in the context of the Pashtun society, where the principle of hospitality (*melmastia*) – which includes spending on social ceremonies – is not only more than just a matter of politeness, but also rather a duty to be fulfilled at any cost, as respectability and social status are crucially shaped by adherence to this custom (Barth 1959; Lindholm 1996). Remittances are hence invested and converted into actions aimed at competing for social status, and accordingly, after investment in land and new houses (35%), in businesses (28%), and social ceremonies and pilgrimage (17% and 9%, respectively) are the most common investment of remittances.

Only for 7 per cent of the surveyed households' remittances were reported to not have significantly changed the lifestyle of the family; yet, this might correspond with the fraction of migrant households that are not remitting as they are still paying back the costs of migration. The repayment of loans that were taken to cover the costs of migration is, for obvious reasons, the first priority and it normally takes from one to two years (Suleri and Savage 2006). In the early 1980s, the total cost involved in reaching Saudi Arabia through an agent was around PKR 50,000, whereas today the figure has peaked to around PKR 470,000, with significant variations depending on the type requested (see Table 6.1).

Despite being a well-established and four-decade-old phenomenon, migrants from KP to the Gulf are organized on the basis of regional committees, whose scope of intervention is limited to extreme

Table 6.1 Costs of migration to Saudi Arabia

Type of visa	Costs (PKR)	Features
Bonded visa	471,600 (ca. 3,600 Euro)	Migrant is bonded to work with the recruiter
Azad (free) visa	628,800	Migrant can work for any employer
Driver visa	733,600	More requested for good status and working conditions

Source: Interview with local recruiter agencies

emergencies (such as collecting money to ship home the dead body of a compatriot) or to serious legal issues faced by Pakistani migrants in the Gulf where migrants are abused and cheated on a regular basis (see Nichols 2008). The *kafala* system at place is highly exploitative and does not ensure basic rights for the workers. Migrants reported long working hours and delay in payments. The fact that the Gulf countries – especially Saudi Arabia – make it basically impossible for migrant workers to live in dignity, to acquire citizenship and to bring their families over must be considered when praising South-South international circular labour migration as the most virtuous type of mobility (see e.g. Hugo 2004). This is also the main reason why NELM as a theoretical framework (and its often criticized focus on male circular labour migration) is well suited to describe migration in the study area. However, there are also dire gendered consequences that need to be addressed and properly investigated. The long absence of male migrant relatives hampered women's mobility and access to health or market facilities, for instance. In the study area, remittances cannot be accompanied by women's increased engagement in work-related activities, as this would entail a loss of respectability (Lefebvre 1999; Siegmann 2010). The social status gained by migrant-sending household coupled with the technological improvement brought by remittances (such as the introduction of less human labour–intensive agricultural technology) has led to a reduction of women's work and exposure to the world. This functions as a symbol of the increased honour gained by the household and is capitalised in what has been aptly defined as 'symbolic capital of honour' (Thieme and Siegmann 2010) to be assigned to women's control and confinement. This is why the reshuffling of male-type task to other males is preferred, and those families where male workforce is no longer available often opt for leasing out farmland to tenants rather than having women involved and thus jeopardize the prestige of the family (Siegmann 2010).

The major Pakistani political parties have representative bodies in the Gulf, but their activity is limited to raising money for the party. Religious channels are preferred when migrant workers intend to donate money for charity. Thirty-two per cent of migrants sending remittances were found to not contribute to their community, 31 per cent would donate to local mosques and madrassas, while 34 per cent would invest individually in charity and poverty relief.

3. Remittances in times of crisis

Remittances and violent conflict

Swat and Lower Dir witnessed a drastic escalation in incidents of terrorism during the last decade. The Talibanisation of the study area peaked in 2007, causing bombings of public infrastructures (schools, places of worship and hospitals), targeted killings and kidnapping and widespread violence (United Nations Development Program [UNDP] 2012). The Malakand Agency and particularly the Swat District together with the Federally Administered Tribal Area bear the biggest brunt of the insurgency that led to a severe conflict costing thousands of lives. The military action resulted in a mass movement of population within different parts of the province. More than two million people fled from the Swat, Lower Dir, Buner and Shangla districts. People were forced to leave their homes, fields and livestock in fear of the militants, who burnt the ripened crops, destroyed orchards, slaughtered livestock and damaged fish ponds as well as the irrigation water channels in almost all areas of the district of Swat. Around 80 per cent of the Internally Displaced Persons (IDPs) have taken shelter with local families or in rented accommodations paid mostly thanks to the money remitted by migrant relatives (Jahangir 2009). Scant attention has been devoted to how local people, particularly in Swabi and Mardan districts, have opened their houses to those fleeing conflict. As put by the chairperson of the Human Rights Commission of Pakistan, Asma Jahangir: 'Had the citizens not acted in a prompt and generous way, protection for the IDPs would have become virtually impossible.' This constitutes quite a unique case, and the reasons for such collective behaviour have to be found in the principle mentioned earlier of *melmastia* as well as in tribal ties, so that people in different districts felt compelled to grant hospitality to fellow IDPs belonging to the same (Yusufzai) tribe. Mobilization from the secular Awami National Party has played a role too.

Remittances were also key in helping temporary displaced people to move back. According to official statistics (UN-OCHA 2010), from

July 2009, the majority (90%) of those IDPs returned to their areas of origin. This is largely due to the above outlined special conditions of hospitality and to remittances inflow. Interviewees reported that 99 per cent of the IDPs have now returned home with the exception of those few who remained in the area of displacement as they started a business there or because their children are now enrolled in schools in the destination locality.

Remitted money played a vital role during the active conflict, as it remained the most important and often the only supply line to ensure the survival of households. Those who remained, due to prolonged and continued curfews and to the impossibility of reaching the market area or to harvest the field and the orchards, were facing starvation that could be avoided only thanks to remittances. Banks were closed for more than four months in Swat and *hundi* was the most effective way of receiving money.

According to the UNDP, approximately 30–40 per cent of the surveyed people in the Malakand Agency are working abroad, and during the insurgency the money sent by these workers ensured the survival of the relatives back home (UNDP 2012). In our sample, 74 per cent of the respondents reported to have avoided starvation during the conflict thanks to remittances, and 9 per cent were able to evacuate the family and rent houses in other cities where their children could also continue their education.

Under this respect, migrant remittances have helped to avoid further displacement, economically supporting people who chose not to leave or could not leave. Those who stay are usually neglected, as they are omitted in the categorizations at place focusing on short- and long-term displacement; yet, they are the most vulnerable and needy (Black et al. 2012).

In the aftermath of the conflict, remittances were a bulwark in the reconstruction of livelihoods and household assets for 23 per cent of the sample. In particular, most of the respondents reported to have used remittances to rebuild their houses. This correlates with recent literature on the Swat crisis, where 59 per cent of reconstruction cost was found to be borne by the houseowner himself, thanks to remittances sent by the relatives working abroad (UNDP 2012). The other major supporter in reconstruction of houses was the Government of KP, which provided funds to 17 per cent of the affected for housing. International donor agencies also supported housing reconstruction in the area (17% of the total reconstructed houses). Relatives of affected households living in non-affected parts of KP and in other provinces also extended their support in housing reconstruction; the share of such families is 10 per cent (UNDP 2012).

Thanks to the complex tribal network of kinship and extended household system, the impacts of remittances were not limited to the recipient household. Often times, non-migrant-sending households also could benefit from remittances that proved crucial in the recovery and stabilization of the local economy and rehabilitation of services during the post-conflict time. FGDs' participants in both Swat and Lower Dir consider remittances as the main reason why recovery has been rather swift in the region. Immediately after the conflict, businesses swiftly reopened and reconstruction was carried out mostly thanks to efforts by private citizens. Some migrants also came back immediately after the conflict, with the precise goal of starting a business and helping reconstruction.

For instance, a participant to the FGD in Swat reported to have migrated to the UAE 22 years ago on a legal visa as labourer. He improved his condition over time shifting from being a labourer to working in a shop. At the beginning, the salary he received was very low and not enough to meet the requirements of his households, but he managed to improve over time and started saving. He came back in 2009, during the conflict. At the time of the military operation in Swat, he and his family left Swat thanks to remittances money. After the military operation, he returned to his village and started a business.

Remittances during and after the 2010 flood

Swat and to a lesser extent Lower Dir were severely impacted by the devastating Pakistan's 2010 flood. By 22 July, tens of thousands were immediately displaced, and up to a million more in the following week as flash floods surged through riverbeds and canals (UN-OCHA 2010). Flooding started along major tributaries, overpowered flood barriers and spread through canals, inundating large swaths of farmland (Mustafa and Wrathall 2011). KP bore the biggest brunt in terms of death, being the province with the highest loss (1,156).

While people were still struggling to rehabilitate their livelihoods from the conflict, the July 2010 flood exacerbated the situation by destroying crops, eroding agricultural lands and killing livestock. Around 100,000 acres of soil along the Swat River had washed out. Many roads and all bridges along the Swat River were swept away (Shah 2010). According to the ACTED rapid need assessment conducted in 173 villages of Swat, 286 people lost their lives and 9,450 houses were completely swept away (Shah 2010). The displacement was localized and involved a large number of people; however, as is often the case, it did not result in long-term displacement (Mueller et al. 2014), as most of the IDPs attempted and succeeded to return to their communities.

As reported, migrant households were able to absorb the economic shock much better than non-migrant households, as they did not depend entirely on income from agriculture and livestock. Seventy per cent reported to have been able to avoid starvation thanks to remittances. They were also able to rebuild the households' livelihood quicker and to buy the households' assets destroyed by the flood (20%). They improved or reconstructed their houses, switching from *kacha* to *pakka* houses (10%). Only 2 per cent of the respondents reported to have evacuated the family in the aftermath of the flood. As tenants are often barred from constructing more permanent or *pakka* houses, the flood provided an excuse for some of them for turning their precarious living arrangements into something less fragile, thus improving their status in the village and their vulnerability vis-à-vis their landlords.

4. Permanent emergency and resilient population: migrants' subjectivities in the making

Migration to the Gulf is entrenched in the society of Swat and Lower Dir to such a point that neither the hard phase of the conflict with the Taliban (2009) nor big-scale environmental hazards like the 2010 flood are perceived to have played a direct role in the decision to migrate by the local people. The large share of the respondents (94%) identified 'Poverty/lack of economic opportunities in situ' as the only and main determinant for migration. 'Gaining social status' and 'environmental hazards' were singled out only by 2.7 per cent of the sample, while a negligible 0.5 per cent indicated conflict as a driver for migration.

The role of remittances is paramount in the communities of Swat and Lower Dir where remitted money acts as a form of household-sponsored insurance system (Yang and Choi 2007). Households shield themselves from risk (including environmental risk ranging from natural weather variability to large-scale disasters) by sending family members to work overseas. The risk might come from the uncertainties over the agricultural production due to climatic and weather vagaries and environmental change as well as from economic losses caused by natural hazards or by violent conflicts. In line with the NELM, and with the understanding of remittances as a household self-insurance mechanism, the local people clearly see migration as the most successful livelihood strategy that allows them to adapt to and/or recover from the effects of potentially hazardous occurrences, be it natural disasters, economic instability or conflict.

In other words, there might be a perceived difference in degree on how pivotal remittances are to guarantee the survival of the household in times of crises as compared to everyday life, but not a difference in nature. Interestingly, during the FGD in Dir, the role of remittances was described as follows:

> Keeping in mind the unstable political and economic situation of the area, remittances work as a bulwark against poverty. During the military operation, most of the people from the Balambat *teh-sil* (Lower Dir, epicentre of the Taliban insurgency) were displaced from their native villages. Those who had remittances money did not suffer, as they could afford renting houses in other cities like Peshawar, Abbottabad, Mardan, Rawalpindi and Islamabad to evacuate their families. Similarly, natural disasters such as the 2010 flood could not reduce migrant sending households to absolute poverty, as disasters affect local resources and economy, but migrants do not depend entirely on them and can survive with remittances. *The true disaster for us would be the drying out of remittances.*

The local perception of the role of remittances in time of crisis is in line with our characterization of labour migration as 'self-insured life', and calls into question the often too rigid and displacement-biased understanding of the relationship between migratory patterns and humanitarian/environmental crises. During the harsh phase of the conflict with the Taliban, remittances improved the survival of people who were cut off from the market and deprived of their local sources of income, and strongly influenced people's displacement behaviours: they allowed for temporary family relocation, they favoured the swift return of IDPs and guaranteed the survival of those who did not displace as well as a rapid reconstruction in the aftermath of the conflict. Also, a similar behaviour was found in response to the 2010 flood. The case study clearly shows that labour migration as a form of self-insurance mechanism has helped in reducing long-term displacement and in ensuring the survival of the household during shock, highlighting the need for better understanding and further enquiry on the interactions between the different forms of human mobility (e.g. displacement and labour migration) that are often part of a continuum (Warner 2009; Beardsley and Hugo 2010).

On the other hand, two important caveat need to be made. First, remittances are generally received by middle- and upper-income families, the poorest usually having lower levels of access to the international labour market (e.g. low level of education, insufficient funds to pay for visa and transport, limited networks abroad; Mazzucato et al. 2008; Le De et al. 2014). In the study area, international labour migration is seen as a painful but rewarding mean of livelihoods and, in the absence of equally available and attracting local opportunities and development, as the most successful livelihood strategy. However, migration is not equally available to all. Besides gender, also ethnicity crucially shapes access to migration. The dominant ethnic group of the region, the Pashtuns, are the most successful migrants, thanks to their higher status, a long story of migration and social and business networks. The socially constructed and embedded narrative on who should be a migrant portrays young Pashtun males as the local 'heroes'.

Increasingly, with the disruption of traditional livelihoods, occupational castes (whose work used to be interdependent to Pashtuns' activities within the village organization) and the Gujjars (originally a nomadic population, engaging in dairy production for the Pashtuns) have recently started undertaking labour migration to the Gulf in order to improve their status and compete with the Pashtuns in the social arena. Yet, access to migration is reserved to non-poorest households. This highlights the need for aid agencies and governments to better reach the most vulnerable segments of society, that is, those who are unable to 'self-insure' their lives through remittances.

There are structural reasons on why vulnerable population in times of crises in the global South (re)produce their subjectivities as self-insured life. A narrow focus on the importance of remittances comes with the danger of obscuring structural socio-economic and political factors, which are the cause of vulnerability in the first place and instrumental to a neo-liberal way of governance promoting 'a post-political life of constant adaptation, [and] the abandonment of long-term expectations' (Duffield 2010: 15; see also Evans and Reid 2013). Remittances indeed help some segments of society (with some beneficial multiplier effects on the community at large) to cope in times of crisis. However, this study also highlights the surveyed communities' structural dependence on international remittances and how deeply socio-political processes (re)produce migrants even before migration starts (see Rodriguez and Schwenken 2013). Portraying migrants as 'heroes of development' (Faist 2008) or 'heroes of adaptation' should not rhyme with leaving the burden of 'adapting', 'developing' and, ultimately, 'surviving' entirely on the shoulders of migrants (de Haas 2010). By preferentially highlighting the agency of migrants, the discourse on remittances in times of crisis normalizes risks and promotes a life of constant struggle for survival, while neglecting and occluding structural root causes of vulnerability. Without proper frameworks and policies at the intersection of development, migration and labour, it is hard to think of migration to the Gulf as a virtuous form of mobility, fostering 'resilience'. Survival – coming at very high costs – seems like a more appropriate term.

Conclusions

Labour migration in the study area is a long-established practice and is well entrenched in the social and economic life of its people. Migration is a strategy to maximize the household income potential and minimize the risks through the diversification of livelihoods in the face of limited local resources, unrelenting overpopulation and a narrow economic base.

The immediate impact of remittances on the household is the improved consumption power in terms of satisfying basic needs, spending and investment in basic amenities, health and education. Therefore, the finding of the study shows an overall positive effect on the living standard of migrants' households. Furthermore, remittances also play a substantial role in increasing children enrolment to schools for both genders.[7] The rapidly expanding private schools are also capitalizing on remittances money that catered co-education, thus helping in bridging the gender education gap. However, the experience of migration is often also the drive behind the demand for better education, as male children are sent to schools having in mind a better job for them as future migrant workers.

When the magnitude is enough, remittances are invested in better housing and business activities as well as in social ceremonies (e.g. pilgrimage) aimed at raising the social status of the household. In hazardous occurrences like natural disasters and violent conflicts, remittances are a fundamental coping mechanism to ensure survival during the emergency and act as an economic stabilizer for the household as well as for the community.

Remittances work as an insurance mechanism with significant impacts. During the harsh phase of the conflict with the Taliban, remittances improved the survival of people who were cut off from the market and strongly influenced people's migratory patterns: they allowed for temporary family relocation, they favoured the swift return of IDPs and guaranteed the survival of those who did not displace as well as a rapid reconstruction in the aftermath of the conflict. A similar behaviour was found also in response to the 2010 flood. The case study clearly shows that proactive migration (labour migration as a livelihood diversification strategy) has helped in reducing long-term displacement and in ensuring the survival of the household during shock. Yet, in the case of both flood and conflict, remittances seldom reach out to the poorest – the most vulnerable segments of society. This shows the need for better understanding of the linkages between different forms of human mobility and draws specific attention to those who are unable to move and have no access to remittances. Praising the role of remittances as self-insurance mechanism and self-help from below should not stop at a mere recognition of their pivotal role. Migration is also undertaken ex-post by disaster-affected households as a way to cope with impacts and losses, and it is often precluded to the poorest, most affected household.

Besides the ad hoc policy for facilitating remittances inflows and reducing the costs of legal transfers of money during a crisis, they should also feature in long-term recovery plan and be taken into account as an income stream for the household (cf Le De et al. 2014). For instance, remittances-receiving households can be provided with loans for housing construction/improvement. In Senegal, thanks to a loan scheme targeting households receiving remittances from Italy, more than 100 families could obtain a loan without a mortgage guarantee and started saving money from the remittances received (IFAD 2013).

This said, the case study also highlights an unhealthy degree of dependency on remittances. As put by one interviewee, the true disaster, for the community of Swat and Lower Dir, would be to not receive remittances. Researcher and policymakers have to be very careful in overpraising the extent to which remittances reduce poverty or build 'resilience'. Remittances in times of crises (environmental or conflict) are crucially important. However, they highlight a harsh struggle for survival, where non-insured lives pay immense human and social costs (and invest considerable financial capital) to barely insure themselves in a precarious life of constant risk and struggle. Can this technocratic, depoliticizing narrative be enough vis-à-vis increasing environmental risk and terrorism?

Notes

1 Acknowledgements: A previous, shorter and substantially different version of this chapter was published: Giovanna Gioli, Talimand Khan and Jurgen Scheffran, 'Remittances and Community Resilience to Conflict and Environmental Hazards in North-Western Pakistan' in Daivi Rodima-Taylor (ed.), *Remittance Flows to Post-Conflict States: Perspectives on Human Security and Development*, pp. 117–126 (Boston: Boston University Creative

Services, 2013). Also available at: www.bu.edu/pardee/files/2013/10/Pardee-CFLP-Remittances-TF-Report.pdf

The author is thankful to the organizers and participants of the COST Workshop 'Climate Change, Migration, Neoliberalism' held in Lund, Sweden, on 11–12 September 2014 for their precious comments and suggestions. My gratitude goes to Mr Talimand Khan, without whom this work would not have been possible.

2 www.iisd.org/publications/environment-conflict-and-peacebuilding-iisd-addressing-links-among-environmental-change, last retrieved 24 June 2015.

3 Some studies also show that remittances can be pro-cyclical, because migrants' decision to remit is also driven by factors such as investment in physical and human capital (e.g. Yang 2008).

4 *Atehsil* is an administrative division of some countries of South Asia. It is an area of land with a city or town that serves as its administrative centre.

5 *Hawala*, meaning 'transfer' (also known as *hundi*), is an informal system for remitting money, in which money is transferred via a network of *hawala* brokers or *hawaladars*. In a nutshell, customer A situated in country X needs to transfer a certain amount of money to customer B in country Y. Customer A contacts a broker in country X and he gives him the amount of money he wants to transfer plus a small commission. Broker A informs a broker in country Y that he has received the money and broker B delivers the amount to customer B after a security check. This is why *hawala* is known as 'money transfer without money movement' (see Ballard 2003). *Hawala* plays an extremely salient role in Pakistan as it presents many advantages as compared to formal value transfer systems, including swiftness, easy accessibility and freedom from bureaucratic procedures as well as cost-effectiveness. Higher trust is also a factor as *hawala* is often accessed via family or social networks and is embedded within the Islamic tradition. On the other hand, the anonymity and efficiency of *hawala* may facilitate the channelling of money for illegal operations, including smuggling, money laundering, tax evasion and terrorism (Fagen and Bump 2006).

6 Telephonic interview with Ahmad Shah, Chairman, Executive Council of Private Schools Management, Swat, 24 April 2013.

7 See Ali (2013) on this issue.

References

Addleton, J. 1984. The impact of international migration on economic development in Pakistan. *Asian Survey* 24: 574–596.

Ahmed, A. 1981. *Dubai Chalo*: Problems in the ethnic encounter between Middle East and South Asian Muslim societies. *Asian Affairs* 12(2): 167–172.

Ali, T. 2013. *Swatis love their schools*. Available at: http://tahirkatlang.wordpress.com/tag/swat/

Amjad, R., Arif, G. M., & Irfan, M. 2012. *Explaining the Ten-Fold Increase in Remittances to Pakistan*. PIDE Working Paper 86. Islamabad: Pakistan Institute of Development Economics.

Arango, J. 2000. Explaining migration: A critical view. *International Social Science Journal* 52: 283–296.

Arif, G. M. 2009. *Economic and Social Impacts of Remittances on House-holds: The Case of Pakistani Migrants Working in Saudi Arabia.* International Organization for Migration (IOM).

Asian Development Bank. 2012. *Addressing Climate Change and Migration in Asia and the Pacific.* Mandaluyong City, Philippines: Asian Development Bank.

Ballard, R. 2003. *A background report on the operation of informal value transfer system (Hawala).* Available at: www.art.man.ac.uk/CASAS/pdfpapers/hawala.pdf

Barth, F. 1959. *Political Leadership among Swat Pathans.* London: Athlone Press.

Beardsley, D. K. & Hugo, G. J. 2010. Migration and climate change: Examining thresholds of change to guide effective adaptation decision-making. *Population and Environment* 32(2–3): 238–262.

Bettini, G. 2014. Climate migration as an adaption strategy: De-securitizing climate-induced migration or making the unruly governable? *Critical Studies on Security* 2(2): 180–195. doi: 10.1080/21624887.2014.909225

Bettini, G. & Gioli, G. 2016. Waltz with development: Insights on the developmentalization of climate-induced migration. *Migration and Development* 5(2): 171–189.

Bisogno, M. & Chong, A. 2002. Poverty and inequality in Bosnia and Herzegovina after the civil war. *World Development* 30(1): 61–75.

Black, R., Adger, W. N., Arnell, N. W. et al. 2012. Migration, immobility and displacement outcomes following extreme events. *Environmental Science and Policy* 27(1): 532–543.

Black, R., Kniveton, D., Skeldon, R., Coppard, D., Murata, A., & Schmidt-Verkerk, K. 2008. *Demographics and Climate Change: Future Trends and Their Policy Implications for Migration.* DFID Working Paper. Brighton: Development Research Centre on Migration, Globalisation and Poverty.

Black, R., Stephen, R. G., Bennett, Sandy M., & John R. Beddington. 2011. Climate change: Migration as adaptation. *Nature* 478: 447–449.

de Haas, H. 2010. Migration and development: A theoretical perspective. *International Migration Review* 44: 227–264.

Dorantes, C. & Pozo, S. 2006. Remittances as insurance: Evidence from Mexican immigrants. *Journal of Population Economics* 19: 227–254.

Duffield, M. 2010. The liberal way of development and the development – security impasse: Exploring the global life chance divide. *Security Dialogue* 41–53. doi: 10.1177/0967010609357042

Ericksen, P., Thornton, P., Notenbaert, A., Cramer, L., Jones, P., & Herrero, M. 2011. *Mapping Hotspots of Climate Change and Food Insecurity in the Global Tropics.* CCAFS Report 5. Copenhagen: CCAFS.

Evans, B. & Reid, J. 2013. Dangerously exposed: The life and death of the resilient subject. *Resilience* 1: 1–16.

Fagen & Bump. 2006. *Remittances in conflict and crises: How remittances sustain livelihoods in war, crises and transition to peace,* International Peace Academy. Available at: www.ipacademy.org/Programs/Research/Prog ReseSecDev_Pub.htm

Faist, T. 2008. Migrants as transnational development agents: An inquiry into the newest round of the migration–development nexus. *Population, Space and Place* 14: 21–42.

FAO. 2012. *Enhancing resilience to food insecurity among protracted crises.* Available at: www.fao.org/ . . . /Enhancing_Resilience_FoodInsecurity-TANGO.pdf

Felli, R. 2013. Managing climate insecurity by ensuring continuous capital accumulation: 'Climate Refugees' and 'Climate Migrants'. *New Political Economy* 18(3): 337–363. doi: 10.1080/13563467.2012.687716

Foresight. 2011. *Migration and Global Environmental Change. Final Project Report.* London: Government Office for Science.

Gazdar, H. 2003. A review of migration issues in Pakistan. Paper presented at the Regional Conference on Migration, Development and Pro-Poor Policy Choices in Asia, June 22–24, Dhaka, Bangladesh. Available at: www.eldis.org/vfile/upload/1/document/0903/Dhaka_CP_4.pdf

Government of Pakistan. 2013. *Migration Statistics of Pakistan.* Islamabad: Bureau of Emigration and Overseas Employment. Available at: www.beoe.gov.pk/migrationstatistics.asp

Hagen-Zanker, J., Mallet, R., Ghimire, A., Shah, Q. A., Upreti, B., & Abbas, H. 2014. *Migration from the margins: Mobility, vulnerability, and inevitability in mid-western Nepal and north-western Pakistan.* Report 5, Overseas Development Institute, ODI.

Higgins, D., Hysenbegasi, A., & Pozo, S. 2004. Exchange-rate uncertainty and workers' remittances. *Applied Financial Economics* 4: 403–411.

Hugo, G. 2004. Asian Experiences with Remittances. In D. F. Terry & S. R. Wilson (Eds.), *Beyond Small Change: Making Migrant Remittances Count.* Washington, DC: Inter-American Development Bank.

IFAD. 2013. Annual Report 2013. International Fund for Agricultural Development. Rome, Italy.

IISD. 2010. *The State of Sustainability Initiatives Review: Sustainability and Transparency.* Winnipeg, Canada: International Institute for Sustainable Development.

IOM. 2014. *Human Mobility in the Context of Climate Change.* Geneva: IOM.

IPCC. 2007. *Contribution of Working Group II to the Fourth Assessment Report of the Intergovernmental Panel on Climate Change.* M. L. Parry, O. F. Canziani, J. P. Palutikof, P. J. van der Linden, & C. E. Hanson (Eds.). Cambridge, UK and New York, NY, USA: Cambridge University Press.

Jahangir, A. (2009). *A Tragedy of Errors and Cover-Ups: The IDPs and Outcome of Military Actions in FATA and Malakand Division.* Islamabad: Human Rights Commission of Pakistan. Available at: http://hrcpblog.wordpress.com/2009/06/04/a-tragedy-of-errors-and-cover-ups-the-idps-and-outcome-of-military-actions-in-fata-and-malakand-division/

Kapur, D. 2005. Remittances: The New Development Mantra? In S. M. Maimbo & D. Ratha (Eds.), *Remittances: Development Impact and Future Prospects.* Washington, DC: The World Bank.

Khan, A. & Khalid, U. 2011. Is Consumption Pattern Homogeneous in Pakistan? Evidence from PSLM 2007–08'. Paper presented at the 27th

Conference of the Pakistan Society of Development Economists, 13–15 December, Islamabad, Pakistan.

Kock, Udo and Sun, Yan. 2011. Remittances in Pakistan: Why have they gone up and why are not they coming down. International Monetary Fund Working Paper No.11/2000. Washington.

Le De, L., Gaillard, J., & Friesen, W. 2013. Remittances and disaster: A review. *International Journal of Disaster Risk Reduction* 4: 34–43. doi: 10.1016/j. ijdrr.2013.03.007

Le De, L., Gaillard, J., Friesen, W., & Smith, F. M. 2014. Remittances in the face of disasters: A case study of rural Samoa. *Environment, Development and Sustainability*, 6(4): 1–20.

Lefebvre, A. 1999. *Kinship, Honour and Money in Rural Pakistan: Subsistence Economy and the Effects of International Migration*. London: Curzon.

Lindholm, C. 1996. *Frontiers Perspective: Essay in Comparative Anthropology*. Oxford-Karachi: Oxford University Press.

Massey, D. S., Arango, J., Hugo, G., Kouaouci, A., Pellegrino, A., & Taylor, J. E. 1998. *Worlds in Motion, Understanding International Migration at the End of the Millennium*. Oxford: Clarendon Press.

Mazzucato, V., Van Den Boom, B., & Nsowah-Nuamah, N. N. 2008. Remittances in Ghana: Origin, destination and issues of measurement. *International Migration* 46: 103–122. doi: 10.1111/j.1468-2435.2008.00438.x

Mohapatra, S., Joseph, G., & Ratha, D. 2009. *Remittances and natural disasters: Ex-post response and contribution to ex-ante preparedness*. World Bank Policy Research Paper 4972. doi: 10.1596/1813-9450-4972

Mueller, V., Gray, C., & Kosec, C. 2014. Heat stress increases long-term human migration in rural Pakistan. *Nature Climate Change*, 4: 182–185. doi: 10.1038/NCLIMATE2103

Mustafa, D. & Wrathall, D. 2011. Indus basin floods of 2010: Souring of a Faustian bargain? *Water Alternatives* 4(1): 72–85.

Myers, N. 2002. Environmental refugees: A growing phenomenon of the 21st century. *Philosophical Transactions of the Royal Society* 357: 609–613.

Nichols, R. 2008. *A history of Pashtun migration, 1775–2006*. Oxford: Oxford University Press.

Paulson, A. 2003. *Insurance Motives for Migration: Evidence from Thailand*. Mimeo, Evanston, IL: Kellogg School of Management, Northwestern University.

Ratha, D. 2003. Workers Remittances: An Important and Stable Source of Development Finance. In T. W. Bank (Ed.), *Global Development Finance*. Washington, DC: The World Bank, pp. 157–175.

Rodriguez, R. M. & Schwenken, H. 2013. Becoming a migrant at home: Subjectivation processes in migrant sending countries prior to departure. *Population, Space and Place* 19(4): 375–388.

Schrieder, G. & Knerr, B. 2000. Labour migration as a social security mechanism for smallholder households in Sub-Saharan Africa: The case of Cameroon. *Oxford Development Studies* 28: 223–236. doi: 10.1080/713688309

Shah, A. 2010. *Household's livelihood trajectories in the context of man-made and natural disasters: A case study from Swat, Pakistan*. Available at: http:// stud.epsilon.slu.se/3664/1/Shah_a_20111203.pdf

Siddiqui, R. & Mahmood, N. 2005. *The Contribution of Workers' Remittances to Economic Growth in Pakistan.* Research Report 187. Islamabad: Pakistan Institute of Development Economics.

Siegmann, K. A. 2010. Strengthening whom? The role of international migration for women and men in Northern Pakistan. *Progress in Development Studies* 10: 345–356.

Siegmann, K. A. & Sadaf, T. 2006. Gendered Livelihood Assets and Workloads in Pakistan's North-West Frontier Province (NWFP). In SDPI (Ed.), *Troubled Times, Sustainable Development and Governance in the Age of Extremes.* Karachi: SDPI, pp. 25–43.

Skeldon, R. 2007. Migration and labour markets in Asia and Europe. *Asian and Pacific Migration Journal* 16(3): 425–441. ISSN 0177–1968

Sriskandarajah, D. 2002. The migration-development nexus: Sri Lanka case study. *International Migration* 40(5): 283–307.

Stark, O. & Bloom, D. E. 1985. The new economics of labor migration. *American Economic Review* 75: 173–178.

Stark, O. & Levhari, D. 1982. On migration and risk in LDCs. *Economic Development and Cultural Change* 31(1): 191–196.

State Bank of Pakistan. 2012. *The State of Pakistan's Economy: Third Quarterly Report for the Year 2011.* Karachi: State Bank of Pakistan. Available at: www.sbp.org.pk/reports/quarterly/index.htm

Steimann, B. 2005. *Livelihood Strategies in Northwest Pakistan: Results from the Sustainable Livelihoods Survey 2004, North-West Frontier Province (Pakistan).* NCCR IP6 Working Paper No. 5. Bern: NCCR.

Stern, N. 2007. *The Economics of Climate Change: The Stern Review.* Cambridge: Cambridge University Press.

Suhrke, A. 1994. Environmental degradation and population flows. *Journal of International Affairs* 47(2): 473–496.

Suleri, A. & Savage, K. 2006. *Remittances in Crises: A Case Study from Pakistan.* HPG Background Paper. London: ODI.

Tacoli, C. & Mabala, R. 2010. Exploring mobility and migration in the context of rural-urban linkages: Why gender and generation matter. *Environment and Urbanization* 22: 389–396.

Tanzler, Dennis, Mohns, Till, & Ziegenhagen, Katherina. 2013. *Adaptation to Climate Change for Peace and Stability.* Berlin: Federal Environment Agency.

Taylor, J. E. & Martin, P. L. 2001. Human Capital: Migration and Rural Population Change. In B. Gardener & G. Rausser (Eds.), *Handbook of Agricultural Economics, Volume IM.* Amsterdam: Elsevier, pp. 457–511.

Thieme, S. & Siegmann, K. A. 2010. Coping on women's back: Social capital-vulnerability links through a gender lens. *Current Sociology* 58(5): 715–737.

UNDP. 1994. *Human Development Report 1994: New Dimensions of Human Security.* United Nations Development Program, United Nations, New York.

UNDP. 2012. *Effects of Militancy and Impact Trends of Rehabilitation in the Malakand Division.* Islamabad: SDPI. Available at: www.sdpi.org/research_programme/researchproject-50-31-562.html

UN-OCHA (United Nations Office for the Coordination of the Humanitarian Affairs). 2010. Annual Report 2009. New York: United Nations.

Vivekananda, J. 2011. *Conflict-Sensitive Responses to Climate Change in South Asia*. London: International Alert, abrufbar unter. Available at: www.ifp-ew.eu/pdf/1111sasia.pdf

Warner, K. 2009. Global environmental change and migration: Governance challenges. *Global Environmental Change* 20: 402–413.

Watkins, F. 2003. "Save There, Eat Here": Migrants, Households and Community Identity among Pakhtuns in Northern Pakistan. In K. Gardner & F. Osella (Eds.), *Migration, Modernity and Social Transformation in South Asia*. London and New Delhi: Sage Publications.

Wisner, B. & Luce, H. R. 1993. Disaster vulnerability: Scale, power and daily life. *Geo Journal* 30: 127–140. doi: 10.1007/BF00808129

World Bank. 2006. *Global economic prospect: Economic implication of remittances and migration 2006*. Available at: http://www.wds.worldbank.org/servlet/WDSContentServer/WDSP/IB/2005/11/14/000112742_20051114174928/Rendered/PDF/343200GEP02006.pdf

World Bank. 2012. *Migration and development policy brief #9*. Available at: http://siteresources.worldbank.org/INTPROSPECTS/Resources/334934-1288990760745/MigrationDevelopmentBrief19.pdf

The World Bank. 2013. Migration and Development Brief. Tech. Rep. 21, The World Bank.

World Bank. 2014. *Migration and Remittances Factbook 2014*. Washington, DC: World Bank.

Yang, D. 2008. International Migration, Remittances and Household Investment: Evidence from Philippines Migrant Exchange Rate Shocks. *The Economic Journal* 118(528): 591–630.

Yang, D. & Choi, H. J. 2007. Are remittances insurance? Evidence from rainfall shocks in the Philippines. *The World Bank Economic Review* 21: 219–248. doi: 10.1093/wber/lhm003

7 Situating migration in planned and autonomous adaptation practices to climate change in Bangladesh

Tasneem Siddiqui, Mohammad Rashed Alam Bhuiyan, Dominic Kniveton, Richard Black, Md. Towheedul Islam and Maxmilan Martin

In the mainstream policy discourse of Bangladesh, migration is generally viewed as a negative outcome of climate change (NAPA, 2005, 2006; BCCSAP, 2008, 2009). Climate change–related migration is perceived to have the potential to compromise recent economic and social achievements of the country. Such understanding led almost all the actors, that is, the Government of Bangladesh, its international development partners and also the non-governmental organisations (NGOs), to concentrate on infrastructure interventions and local-level livelihood adaptation programmes to reduce the need for migration. This paper demonstrates that contrary to planned adaptation programmes, the autonomous adaptation practices of affected households have integrated different forms of migration as one of the many adaptation strategies. It also establishes that adaptation outcome of migration is context specific and under some social, economic, demographic and policy environment migration that leads to adaptation while in some other situation it leads to maladaptation.

This chapter used both primary and secondary data generated in 2013 by RMMRU and SCMR to understand climate change–related migration in Bangladesh. Primary data combined both qualitative and quantitative approaches. Qualitative approach included a focus group discussion (14), village-level community surveys (14) and in-depth interviews (25). Three regions, all hotspots of environmental and climate change hazards, formed the basis of this empirical research. Chapai Nawabganj suffers from drought, Satkhira experiences cyclones and saline intrusion and Munshiganj faces floods and riverbank erosion, which are also both common in the previous two areas.

Table 7.1 Geographic distribution of sample households by gender

District	Chapai Nawabganj	Dhaka	Satkhira	Khulna	Munshiganj	Keraniganj	Grand Total
Male	214 (70.9%)	183 (91.5%)	168 (56%)	120 (60%)	160 (51.9%)	171 (89.1%)	1,016 (67.6%)
Female	88 (29.1%)	17 (8.5%)	132 (44%)	80 (40%)	148 (48.1%)	21 (10.9%)	486 (32.4%)
Total	302	200	300	200	308	192	1,502

Source: All figures and tables in the chapter have been sourced from 'Climate change related migration in Bangladesh'; RMMRU and SCMR; 2014

Table 7.2 Geographic distribution of sample households by migration experience

District	Chapai Nawabganj	Dhaka	Satkhira	Khulna	Munshiganj	Keraniganj	Grand Total
Number of Migrants	118 (39.1%)	200 (100%)	100 (33.3%)	200 (100%)	52 (16.9%)	192 (100%)	862 (57.4%)
Number of Non-Migrants	184 (60.9%)	0	200 (66.7%)	0	256 (83.1%)	0	640 (43.6%)
Total	302	200	300	200	308	192	1,502

Quantitative approach included a longitudinal survey of 862 migrant and 640 non-migrant households spread over these three origin districts as well as two destination districts: Dhaka and Khulna. These 1,502 respondents were selected from a larger data set of 3,000 households generated by a Rapid Screening Survey (RSS). In origin areas, RSS used three criteria for selection: age, gender and migration experience. In destination areas, the district of origin was added to the previous criteria (see Tables 7.1 and 7.2). In addition, this paper drew secondary data from a literature review focused on climate change, migration and government policies relevant to these issues.

Definitions used

Migration is defined in this paper as a process by which an individual, household, group or community – voluntarily or involuntarily – leave their usual place of residence for another location within or beyond national borders, either permanently, temporarily or cyclically. Pre-dominant motivations to migrate include the incentive to be closer to opportunities, resources and people or to avoid persecution.

Adaptation to climate change is seen here as a process of adjust-ments that reduce vulnerability to climate variability and change.

Literature review

Recent empirical case studies demonstrate a more nuanced under-standing of the relationship between migration and adaptation, high-lighting that migration is an important tool for environment and climate change adaptation (Hunter et al., 2015; McLeman and Smit, 2006; Tacoli, 2009). Barnett and Webber (2010) demonstrate that by migrating from climate-stressed areas, people can reduce their vulner-ability and ensure access to income sources unlikely to be affected by a disaster. They further note that post-disaster remittances from rela-tives help households to recover from losses. Tacoli (2009) shows that remittances from migrants facilitate agricultural adaptation in vulner-able communities in Bolivia, Senegal and Tanzania, while Ezra (2001) demonstrates that families with access to remittances better adapt to livelihood crises than those with no access to remittances. Moreover, Kothari (2003) notes that migration often helps reduce pressures on local ecology and natural resource dependence. Such migration is labelled as environmentally induced labour migration (Afifi, 2011). Black and colleagues (Foresight, 2012) noted that migration may lead to reduction in vulnerability through enhancement of livelihoods. In some contexts, migration may offer the most direct form of adaptation as it provides opportunity to affected people to move from hazardous locations. Siddiqui and Billah (2014) found that affected households which combined local-level strategies such as switching cultivation practices and construction of dykes among others with migration were more successful in adapting to climate- and environment-related stresses. Larger family units, which remained in origin areas, availed local government and NGO health, education and nutrition services and diversified their income sources through the labour migration of one or more household members to urban areas.

Warner and Afifi (2014) further investigated the complexity of the relationship between migration and adaptation, concluding that it is best expressed in terms of a continuum. According to them, mobility and immobility can be both positive and negative forms of adapta-tion to climate change at the individual and household levels, depend-ing on the household context. They have come up with four types of migration categories, based along a continuum ranging from resilience to vulnerability. 'Adaptive migration', the most resilient category, involves diversification of livelihoods and increased access to education, health and political resources. 'Survival migration' ensures that a household survives but does not flourish. Even more vulnerable are those who participate in 'last-resort' migration, which is based around 'erosive coping strategies' who are driven by hunger and desperation. Scenarios involving trapped populations without any opportunity to migrate constitute the fourth and most vulnerable category on the continuum. Households which have access to livelihood options and social, economic and political assets, whose children have three to five years or more education than their parents and have one or more young members with the ability to migrate and send remittances back to their families belong to the 'adaptive' category. On the other extreme, landless households that suffer from chronic food insecurity and are unable to migrate or provide education to their children, fall under the 'trapped' category.

We can sum up the literature by saying that migration can be one of the many adaptation tools employed by affected families. Not all forms of migration lead to adaptation. Some may lead to maladaptation. However, turning such migration into adaptation requires policy support at both the local and the national levels. This paper discusses these issues at length.

Experiences of environmental and climate change hazards

Both quantitative and qualitative data of the RMMRU and SCMR surveys indicate that most of the people of Satkhira, Munshiganj and Chapai Nawabganj faced all kinds of environmental and climate change–related hazards. Quantitative data shows that out of 1,502 households, 75 per cent including both migrant and non-migrant households faced such stresses in their dwellings in different periods of life. Of those who reported facing environmental or climate change–related stresses, more than half (64%) faced multiple hazards, 12 per cent faced riverbank erosion, 8 per cent faced drought and 6 per cent faced flooding. Households of Satkhira faced cyclone, riverbank and coastal erosion, salinization, reduced crop yield, water logging and flood. Households of Chapai Nawabganj mainly faced drought, lack of rain, erratic rain, declining land fertility, reduced crop yield and seasonal flood. Households of Munshiganj faced flood, riverbank erosion, reduced fish catch and reduced crop yield (Fig. 7.1).

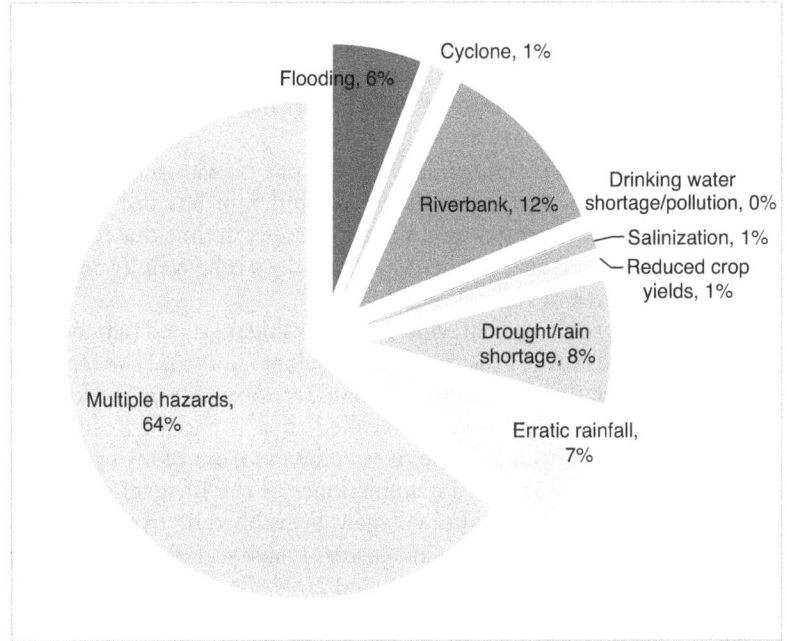

Figure 7.1 Experiences of environmental and climatic hazards

Abul Kalam Fakir (47) of Gorr Kumarpur village, Padmapukur union of Satkhira, stated: 'My family survived cyclone Aila. My homestead is also vulnerable to riverbank erosion. After Aila I cannot grow any crop or plant tree due to saline water intrusion and water logging.'[1]

Shumita Munda, a 25-year-old woman of Gabura village of Satkhira, stated, 'My family has experienced Aila in 2009. I have never seen such kind of heavy tidal flow in my life. It has taken away everything. My house was completely destroyed and all my cash and resources were washed away. After Aila there is hardly any tree in the village, whereas in the past the village was green. Now we are observing unbearable heat. Soil and water have completely become salinized that are not conducive for cultivation.'[2]

The Khesba villagers of Chapai Nawabganj recalled that ground water level was around 60–70 feet in the 1990s. By 2000, it had dropped down to 70–80 feet, and by 2012, it had fallen to 120–140 feet. Approximately 70 per cent of the village land faced severe ramifications because of the drought. Thirty-year-old Salahuddin of that village says, 'I used to farm here with my father and brothers, we used to grow paddy in one season and a Rabi crop in another, now we can't even grow one crop in a year, as the ground water has dropped really low and the rains aren't enough. I don't have any other work and I can't survive as a farmer.'[3]

Zia-ur (40) of Nijampur village of Chapai Nawabganj says, 'The amount of rain in a year has decreased and rain has become more irregular. This year farmers could not sow seeds in time due to lack of water caused both from inability of current deep tube wells for extracting water and lack of rain.'[4]

Golapjan (60) of Nagerhati village of Munshiganj stated that increasingly early floods caused disruptive crop submersion, and she reported also losing some of her land to riverbank erosion. Aleah Sheikh (50) of Kolma village of Munshiganj recalled that 'a part of Kolma and Daidia village was eroded by the river Padma in front of my eyes'.[5] Mr Mahbub Rahman (35), ward commissioner of the Bhagyakul village, says that 'the whole Bhagyakul village was grabbed by river Padma in 2010. Flood intensity has been increasing. I have seen three big floods, one in 1988, and other two in 1998 and 2004. Each time the intensity of flood increased'.[6]

In each of these three areas, evidence of climate change and environmental variability is already a visible part of local experience. Riverbank erosion and floods are common to all areas, besides the manifestation of environmental and climate change that varies according to location.

Adaptation practices in study areas

Earlier, we defined adaptation as adjustments made by households or communities to reduce vulnerability to climate variability and change. People of Sathkhira, Chapai Nawabganj and Munshiganj have been trying to adapt to the risk and vulnerabilities mentioned earlier in different ways. Some were initiated by the government, some by NGOs and some others were developed as autonomous adaptations by respondents themselves. We have divided them under two categories: planned adaptation and autonomous adaptation practices. Planned adaptation practices are mainly initiated by the government, local and international NGOs and development partners. Autonomous adaptations are initiated primarily by the affected people and their community.

Planned adaptation

Interviewees of Satkhira identified Disaster Risk Reduction (DRR)-related infrastructural interventions such as construction of cyclone shelters, construction of polders in the coastal areas and construction of climate resilient homestead as planned adaptation interventions. Different ministries and agencies are mostly involved in infrastructure development activities whereas construction of climate resilient housing is facilitated by local and international NGOs. Shushilon, Rupantor, Oxfam UK, Water Aid, Muslim Aid, Caritas, Save the Children and BRAC are working in Satkhira. Pond dewatering, raising of tube well for ensuring disinfected water supply, Pond Sand Filter (PSF),[7] water purifying tablet distribution, water treatment, regular water quality monitoring, rainwater harvesting, re-excavation of ponds, water conservation for drinking and domestic use, multiple uses of sweet water, canal re-excavation and conservation of rainwater for irrigation, seeds distribution of saline resistant crop (BIRI-45, 47 and 55), aquaculture and crab fattening in saline water, installation of solar electricity and building awareness and disaster preparedness are major forms of intervention by NGOs and INGOS in Satkhira. Some NGOs also provide credit services for accessing solar-based electricity.

Chapai Nawabganj respondents identified three types of government programmes, namely drought management, deep tube well-facilitated irrigation, short duration, drought tolerant seeds distribution and improved storage system of seeds. NGO- and INGO-implemented programmes include water conservation, minipond digging for water conservation and retention of rainwater and moisture in Khari canals, storage of seeds and fodder, crop diversification and intensification; *lakkha* (lacquer) cultivation and planting of *boroi* (sour plum) trees, vegetable gardening and lots more.

In villages of Munshiganj, government or NGO interventions to adapt to challenges of floods and riverbank erosion were more limited than the other two areas. Government intervention largely focused on construction of dams, dredging and river training. At the time this research was conducted, NGOS and INGOs who work on climate change were not active in this area. By contrast, the majority of the NGOs active in this area were implementing micro credit, health and women's empowerment programmes.

While reflecting on the ongoing efforts of planned adaptation, the respondents of all three sites appreciated the construction of different disaster risk-reducing infrastructure and the excavation of rivers, as these interventions have reduced risks of death during cyclone or to continue to use the river for livelihood. However, respondents also highlighted the need for sustainable livelihood options. Most of the programmes do not sufficiently address the requirement of alternative income sources. Due to loss of avenues for earning income, some of the households are falling into occasional or extreme poverty traps. For example, Abdur Razzak (40) of Central Khalishabunia village stated: 'Most of the time I am unemployed. When polders are constructed or maintained I get to work as day labourer. I have received some money from cash for training programme after Aila. However, that was spent on day-to-day expenses of my household.'[8]

Rina Begum (28) from 9 Number Sorra village of Gabura also received cash for training from a local NGO. She stated: 'Three years after Aila, when the rain started, I planted spinach in my backyard. However, I cannot maintain my livelihood on this. I need a continuous source of income. The impoverishment of my family is increasing day by day.'[9]

Merina (35) of Chapai Nawabganj stated: 'Short-duration rice seed is available locally. But my husband has stopped growing paddy. Due to lowering of level of ground water, one needs to invest more in irrigation. We can no longer afford irrigation as it has become extremely expensive. Only the rich can purchase irrigation water and continue to cultivate.'[10]

Autonomous adaptation

Different types of autonomous adaptation practices were mentioned by the respondents. They included small infrastructural innovations, changes that they brought in their lifestyle, changes in livelihood and increased use of various types of migration. The respondents shared examples of some simple changes that they incorporated in their day-to-day life. To cope with the extreme hot temperature during the day, people stay outside their homestead. Shumita Munda (25) and Jahangir (38) of Gabura village of Satkhira stated: 'We locate ourselves under big trees beside the river, ponds and embankment.'[11] Women mentioned that use of hand fans have increased. FGD participants of Satkhira and Chapai Nawabganj mentioned: 'We wear loose cotton attires.'[12] Salahuddin (30) of Khesba village of Chapai Nawabganj stated, 'I have minimized my family food expenditure by changing food habits.'[13] People of Satkhira and Munshiganj avoided purchasing fish, instead consuming only those fish which they caught. Respondents of Satkhira and Chapai Nawabganj mentioned that they work less hours per day during hot summers, and also adjusted the time of day in which they worked. Md. Abu Daud Gazi (45) from Gorr Kumarpur village of Satkhira district stated: 'We begin very early in the morning and stop working when the sun is strong and again begin work in the afternoon.'[14]

We have seen in the earlier discussion of planned adaptation practices that government, national and international NGOs played a major role in constructing DRR infrastructures, in modifying dwelling structures and so on. Along with those, affected communities have also initiated certain action on their own to reduce their vulnerability to hazards based on their traditional knowledge. Respondents of Munshiganj built their houses on pillars raised from the ground level to avoid the submersion of their dwellings during normal floods. This also allows them to dismantle the home structure quickly during riverbank erosion. Autonomous infrastructural adaptation practices of Munshiganj also included construction of bamboo walls beside river beds to control riverbank erosion. A section of villagers have also introduced commercial livestock rearing to create extra income sources. Some of the affected families have started cultivating special types of fodder grasses including Ipil-Ipil (*Leucaena leucocephala*) and Hybrid Napier and more.[15] The commercial livestock rearing has made these grasses very much in demand as they are used as cow fodder.

Most of the major adaptations that the people in these locations practised involved changes of livelihood types. In this section, we have located all types of traditional livelihoods that people have lost and new livelihoods that people autonomously adapted at the local level. We will not discuss new livelihoods which were accessed through migration here. We will do that in the following section on migration as one of the tools for autonomous adaptation. However, we are very aware that not all livelihood losses are due to change in climate or environment; some are related to government policies, landlessness and so on. Yet, it is understood that a large section of the losses are connected to it. Therefore, it is important to identify the extent of livelihood losses in the study areas as a whole.

Qualitative data captures the livelihoods of the respondents over the last 40 years. It shows that before the 1980s, people in Satkhira used to work as subsistence farmers, sharecroppers, agricultural labourers, fishermen, boatmen, woodcutters, honey and leaf collectors, herdsmen, potters and smiths. Members of poorer households used to migrate to other districts seasonally during sowing and harvest times. In the 1990s, shrimp and crab cultivation began in the area. A section of the poor households found work as labourers in these farms.

Shrimp cultivation is less labour intensive, and as part of the forest preservation effort, restrictions were imposed on gathering resources from the Sundarbans. Members of some of the households, who earlier depended on the forest for their livelihoods or who did not possess land or who could not cultivate crops due to water logging or salinization, found themselves working locally as day labourers, shrimp fry collectors, rickshaw/van pullers, passenger-carrying motorcycle drivers and unskilled workers in different sectors, notably fish processing, petty trading, shopkeeping and more.

In Chapai Nawabganj, village livelihoods centred around agriculture, involving seasonal migrations as agricultural labourers as well as cross-border daily commuting (for agricultural work in India). Animal herding and seasonal fruit businesses also constituted major livelihood opportunities for the villagers in the past. A majority of the villagers of Shibganj *upazila* experienced loss of land to riverbank erosion, and all of them experienced drought. Commuting between India and Bangladesh border for work also ceased due to imposition of a strict border control regime. Over the years, agriculture has declined as a primary income source. Day labouring and rickshaw and van pulling along with some agricultural work within the village (or nearby villages) became the major livelihood options. Nonetheless, a large number of working-age men would have found themselves unemployed if they did not migrate.

In Munshiganj, fishing and agriculture were the two major areas of employment in the past. However, the scenario has changed. Now, according to the villagers, agriculture constitutes the primary source of income for only one-fifth of the families. Because of seasonality of income, members of fishing communities also got themselves involved in other types of jobs, which are mostly available outside the village. The size of the fishing community has, therefore, shrunk significantly. People who owned land and water bodies got involved in pisciculture. Some also transformed their paddy fields into ponds in order to engage in pisciculture. Now, pisciculture has become the primary income-earning source of a section of households in Munshiganj. However, it does not create any significant employment for others. This means a large number of people who lost land to riverbank erosion, small land holders, landless fishing community and day labourers would remain unemployed or underemployed in the village if they had not participated in wider labour market through temporary, seasonal or permanent migration.

Though both NGOs and the government have been active in creating programmes to facilitate adaptation to climate change and environmental variability in these three regions, people have also autonomously initiated various adjustments to their own lifestyles and livelihoods to better adapt to these changes.

Use of migration as one of the tools of adaptation

Kniveton and colleagues (2013) developed a future climate change and migration scenario of Bangladesh. Based on historical analyses of *upazila*-level census data from 2001 and 2011, as well as predictions of global climate models and the World Bank Studies of 2010 and 2011, the study projected that from 2011 to 2050, as many as 16–26 million people would migrate from places of origin due to floods, storm surges, riverbank erosion and sea level rise. Two to five million people of this group will migrate due to riverbank erosion. Three to six million will migrate due to inland flooding and five to seven million will migrate due to coastal storm surges, while six to eight million will migrate due to sea level rise.[16] The influence of climate change alone will cause at least 500,000 to 1 million extra migrants at this time scale.

Villagers of our study areas experienced several types of migration. Some had strong elements of forced migration and some others had voluntary elements; every case of migration identified in our study was an autonomous rather than planned government or NGO-led process. We found no examples of planned resettlement of displaced persons in these three areas. We have identified seven distinct types of flows in the study sites. These are: movement of partial or whole villages due to displacement; individual or household self-relocation within nearby villages or towns; temporary settlement in roadside or embankment shelters; permanent migration of relatively wealthy households to safer locations; internal migration of single/multiple family members to support families in origin areas; cross-border/regional movements; and short-term contract migration to Gulf and Southeast Asian countries.

Movement of partial or whole village due to displacement

Char Paka village of Chapai Nawabganj and Bhaggyakul village of Munshiganj were devoured by the river. The river Padma washed away the entire village of Bhaggyakul, which was around 300 hectares in size. Only a fish market now remains, adjacent to Mandra village. Villagers of Bhaggyakul have taken shelter in the nearby villages of Mandra, Charipara, Rarikhal and Khamargaon. Some of them bought land, while others are squatting on land they do not own. Some of these set up informal shelters on government-owned land. Those who could buy land have high likelihood of representing examples of adaptation. However, most of those who are squatting may fall under the category of 'vulnerable', due to their dependence on erosive, 'last resort' coping strategies. Of course, one can pursue rigorous analysis on this by cross-examining such information with age, gender and education level of the respondents by their migration experience.

Two unions, 12 No Paka Union of Ujirpur and Golakpur Union of Shibganj, lost nearly all their land in the 1980s to riverbank erosion. The former unions together constituted 48,000 *bighas* (land), of which 42,000 *bighas* were lost over the last 40 years. Each household of this village experienced the shifting of homesteads on at least four to five occasions in their lifetime due to erosion. Ten FGD participants were subjected to erosion 10–14 times in their lifetimes. In the 1970s, another char emerged out of the river, and many members of the two unions resettled themselves upon this new char forming the Char Paka village. If this group of people were unable to form this new village, they would have been defined as trapped population. However, the natural phenomena of char formation permitted the villagers to autonomously adapt to their environmental stresses. This constitutes an example of successful adaptation through migration. In Warner and colleagues' (2014) conception, this might be considered an illustration of resilience through 'adaptive migration'.

Relocation of homestead

Quantitative data gives us an idea about the number of times each of the surveyed households relocated their homestead. The data shows that migrant households, from both origin and destination areas, on an average had to relocate their homestead 2.78 times. On the other hand, non-migrant households from both origin and destination areas relocated their homestead 2.40 times. Among the three study areas, Munshiganj is the area most prone to riverbank erosion, and thus relocation experience is highest here. Migrants' households of Munshiganj relocated themselves 3.48 times, while non-migrant households relocated slightly less than 3 times.

Respondents identified different reasons for relocating their homestead. Figure 7.2 shows that 30 per cent stated that they relocated their homestead due to climatic stresses, 4 per cent thought climatic stresses partially influenced their relocation decision and 24.35 per cent categorically stated that climatic stresses did not have any bearing on their

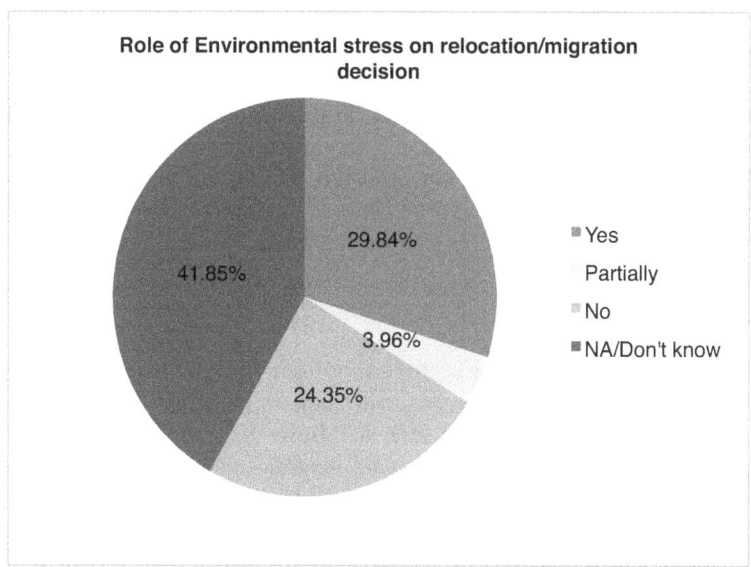

Figure 7.2 Relocation experiences of homestead

relocation decision. As many as 41.86 per cent did not know whether climatic stresses influenced their decisions or not.

Relocation experience highlights that as many as one-third of the households attributed their decision to relocate homestead to climatic stresses. Relocation experiences may represent both autonomous adaptation and maladaptation in the face of climatic stresses.

Permanent migration

Our study identified two streams of permanent migration, occurring particularly from Satkhira. After cyclones Sidr and Aila, 25 per cent of the relatively better-off families migrated permanently or moved their families mostly to nearby Khulna and Jessore districts (FGD). Such families of Munshiganj also migrated to Dhaka and nearby districts autonomously.

FGDs of Satkhira also revealed that after Aila, 150 families of Hindu faith permanently migrated to India. Almost all of the migrating households to India had a section of their family members already residing in different parts of West Bengal. After Aila, it was easier for them to decide to migrate permanently. One of the areas of Chapai

Nawabganj shares a border with India. However, incident of cross-border migration was reported neither from Chapai Nawabganj nor from Munshiganj. Many of the men and women who migrated from Munshiganj, and who are now employed in the formal labour sector (i.e. garment and leather manufacturing), now identify as permanent migrants to Dhaka, Gazipur and Narayanganj. From the perspective of the affected households, the decision to migrate permanently for these formal sector works can be cited as examples of resilient 'adaptive migration'.

Temporary relocation

In Satkhira, after the Aila, a section of households who lost their homesteads resettled themselves in roadside shelters, on embankments, or in houses of relatives residing in safer locations. In Munshiganj, as well, some of the respondents squatted on neighbours' lands and some of them rented dwellings. Initially, almost all tried to move within the village. When it was not possible, they moved to nearby villages. These can again be examples of both 'survival migration' and/or erosive, coping strategies, which, as previously mentioned, can be considered more vulnerable types of adaptation through migration.

Seasonal migration

Rural-to-rural seasonal migration during agricultural season historically represented a traditional livelihood strategy in certain regions of Bangladesh. Among the three study areas, it is currently most common in Satkhira. Men mostly migrate as agricultural workers to different parts of Bangladesh, including the rural areas of Jessore, Jhenaidah, Khulna, Gopalgonj, Madaripur, Pirojpur and Sylhet. The cost of migration ranged from BDT20 to BDT600. The average duration of stay in destination areas was between 1 and 2 months seasonally during sowing and harvest time. Villagers of Satkhira now also migrate to peri-urban areas of different districts such as Gazipur, Munshiganj, Narsingdhi, Narayanganj, Savar (Dhaka), Chittagong, Feni and Faridpur to work in the brick kilns. Their migration season lasts for six months.

Internal migration from Chapai Nawabganj, on the other hand, flows mostly to urban megacities. However, a section of them do migrate seasonally to rural areas of Rajshahi, Bogra, Pabna, Faridpur, Manikganj and so on. The migrants are also mostly men. Seasonal migration does create alternative avenues of income, and thus can be seen as one of the avenues of income adaptation. This seasonal, cyclical migration occurs while families remain in areas of origin; these movements may constitute income diversification strategies, and thus be considered as instances of successful 'adaptive migration'.

Rural-urban temporary labour migration

Rural-urban migration for work is common to all three locations. Major rural-urban migration from Satkhira flows predominantly to Khulna, Jessore, Gopalganj, Barisal, Bhola, Patuakhali and Madaripur. In the destination areas, migrants most commonly work in paddy processing or rice mills, or as labourers in Mongla port and market places, rickshaw and van pullers, scrap collectors, shopkeepers, vendors, traders and more.

The construction boom in cities like Dhaka and, to some extent, Chittagong created a new job opportunity for the migrant households of Chapai Nawabganj. Workers from Gomostapur Upazila and Chapai Nawabganj can be found in most of the construction sites in Dhaka and Chittagong within both the skilled and unskilled groups. Rickshaw pulling is another important profession that provides employment opportunity to the villagers of this district. During the FGD, it was found that 80 per cent of the working-age men of drought-prone Khesba village in Chapai Nawabganj now migrate to Dhaka for rickshaw pulling and construction work.

Destinations of temporary rural-urban migrants of Munshiganj are Dhaka, Gazipur, Narayanganj and Narsingdhi. A section of them work in textiles, garments manufacturing as well as leather, glass and steel factories. From Munshiganj, both men and women migrate for work. Men hailing from relatively better-off families of Munshiganj have migrated to a specific destination in the outskirts of Dhaka city (i.e. Keraniganj). They dominate the marketplaces of Keraniganj, Islampur and Nawabpur. Some are owners and some work as salesman in these markets. Since Mushiganj is close to Dhaka and other industrial districts, some commute there and work as day labourers, hawkers, vendors and rickshaw pullers. Many of these temporary workers migrate as individuals and remit regularly to their families remaining in their origin villages. A small section of these workers gradually bring their members to be with them and transition into permanent migrants. Given these evidences, it is clear that rural-urban migration for work has major potential as an adaptation tool.

Short-term international contract migration

Since the 1980s, some of the villagers of Munshiganj have been migrating to the Gulf and other Arab and Southeast Asian countries as short-term international contract migrant workers. Economic condition of those households whose members have taken part in international migration is much better compared to similar households (class, location etc.) whose members did not. Many of them have been able to construct *pucca* houses in relatively safer locations.

Recent research (Sidddiqui and Mahmood, 2015) has compared the poverty situation of short-term international contract migrant and non-migrant families who initially belonged to similar socio-economic background. It found that 13 per cent of the international migrant households were below the poverty threshold, whereas 40 per cent of the non-migrant households belonged to that category. This illustrates that, in these cases, migration was associated with a relative reduction in poverty. This research covered 17 districts, including Chapai Nawabganj, Satkhira and Munshiganj.

People from Satkhira were negligible participants in the short-term international contract labour market. Traditionally, households of Chapai Nababganj did not have access to international migration. When the United Nations and the United States lifted the sanction on Libya in 2008, Chapai Nababganj became a new source area for recruiting migrant construction workers headed to Libya. However, many of those who migrated to Libya were forced to return to Bangladesh after the onset of the Libyan crisis in 2011. The majority of these incidents represented failures of migration involving substantial loss of property and assets. Therefore, it is clear that not all short-term international contract migration leads to better income and livelihood adaptations. Besides, fraudulent practices in the recruitment of international labour migrants often reduced the potential of income improvement from such migration.

Drivers of migration

It is interesting to note that although the respondents clearly identified the impact of climate change on their lives, only 10 per cent attributed their primary reason of migration to climatic stresses. By contrast, 27 per cent stated that they migrated because they received information regarding available work at destination or because lack of work in their origin area (Table 7.3). Another 21 per cent stated that they migrated because they had no income at all in their origin area, while

Table 7.3 Major drivers of migration in origin and destination districts of Bangladesh*

Reasons for migration	Satkhira	Khulna	Chapai Nawabganj	Metropolitan Dhaka	Munshiganj	Keraniganj (Greater Dhaka)	Percentage of total
Natural Calamities	1.6%	24.0%	0.8%	1.9%	8.1%	24.2%	10.10%
Earn an Income**	71.0%	57.6%	70.8%	89.5%	72.0%	64.5%	70.90%
Business	1.3%	1.1%	5.0%	0.5%	0.5%	3.6%	2.00%
Family Migration	12.9%	10.9%	7.0%	5.5%	8.1%	2.7%	7.85%
Marriage Migration	0.0%	3.4%	6.9%	1.2%	9.7%	2.7%	3.98%
Student Migration	3.5%	2.0%	3.0%	0.0%	0.0%	1.4%	1.65%
Others	9.7%	1.0%	6.5%	1.4%	1.6%	0.9%	3.52%
Total	100.0%	100.0%	100.0%	100.0%	100.0%	100.0%	100.0%

Note: *Each column represents 100 per cent, and thus the figures of each cell are percentage of the column.

**'Earn an Income' includes migration due to poverty, information of availability of work at destination, lack of work in origin area, no income at all in their origin area, some income in the origin area, but migrated with a desire of better life and income.

13.5 per cent reported some income in their origin area but migrated because of desires for a better life and better income. Poverty (9.1%) was also stated as a reason for migration. Accompaniment of migrating family members (7.8%), marriage (4%) and education or studies (1.6%) were also given as reasons for migration.

In the qualitative survey, some linked their migration decision with climate stresses, but others clearly stated that climatic stresses had nothing to do with their decision to migrate. For example, Aslam (30) of Nagerhati village of Munshiganj stated: 'I was unemployed before migration. My migration was driven by my desire to lead a better life not by any natural calamities. I learned the carpentry and started doing business in Rohitpur Bazar. I raised the initial capital by selling a piece of land.'[17]

Golapjan (60) from Munshiganj lost her land to riverbank erosion, and yet she did not frame her daughter's migration as a response to this calamity. Her family lost their agro-based livelihood and was living in extreme poverty, yet she stated: 'International migration of my daughter is not a response to flood or her loss of land to riverbank erosion. It is rather a household income strategy. But it is true that if river erosion did not take place and did not devour my land my family could be well off.'[18]

Monira (23), from Gabura, Shyamnagar, Satkhira, stated: 'When cyclone Aila hit, we lost everything we had. As there was no livelihood opportunities to earn an income here, I moved to Dhaka with the help of my uncle and started to work at a garment factory and become self-sufficient. Now I earn Tk. 4,000 ($50) a month and send money for my family through Bkash. I was one of the first girls from the village to move to Dhaka for job after Aila and now most of my neighbourhood girls want to go Dhaka to get a job as there are still no opportunities to work here.'[19]

Md. Manik (32) of Chapai Nawabganj, however, put more emphasis on livelihood losses due to climatic stresses as a major factor. He says, 'When there is no work in the village particularly during the drought and rainy seasons I work outside of my village. At least for four months.'[20] He does not consider this in terms of migration, but instead frames his movements as short-term work trips.

Traditionally, Sumita Mundas (25) whose family live in Satkhira was involved in farming. 'Due to shrimp cultivation and intrusion of saline water in the agricultural land and the cyclone of 2009 had created water logging in my village for the last 2 years. Agriculture is no longer viable due to saline water intrusion. My husband now works as agricultural labour in other districts.'[21] Aleah Sheikh (50) of

Munshiganj stated: 'River Padma devoured 33 decimal of my lands and I lost my regular income. After that one of my sons migrated to Dhaka and regularly sends remittance.'[22]

In the quantitative survey, when they were asked what was the reason behind their migration, respondents directly linked this decision with income factors such as joblessness, better opportunity and so on; however, during the qualitative survey, half the respondents identified migration as outcome of income loss due to climatic stresses. One interesting trend that arises is the tendency to only treat movements of whole families as 'migration'. By contrast, when family members remained in origin locations while only one or a few members migrated, respondents did not perceive this as migration.

Comparison of responses between migrants in destination and families of migrants still located in origin area

In the survey, 24 per cent of migrants from Satkhira to Khulna and 24.2 per cent from Munshiganj to Dhaka identified climatic stresses as primary reason behind their migration. However, only 1.6 per cent of migrants' family members who remained in Satkhira, an area that experiences major storm surges, attributed their relatives' migration decision to environmental factors. Similarly, in Munshiganj, an area often beset by riverbank erosion, only 8.1 per cent of family members left behind identified environmental factors as the primary reason behind their relative's migration. Of the few (0.8%) of the respondents of Chapai Nawabganj and its migrants in destination (Dhaka), only 1.9 per cent identified climatic stress as reason for migration. This reveals that people from areas that experience rapid-onset climate events are more likely to link their migration with environmental factors or climate change than people from areas that experience slow-onset climate events such as drought. Again, persons who have migrated and are currently staying in the destination, correlate their migration decision more with climate change compared to those members of migrant households who remained in the origin area.

Analysis of interview responses from these five locations highlights that migration experiences are also gendered. Males and females of the interviewee households have different migration experiences. This difference is mostly created by the nature of social network and job market demands of destinations. Most of the migrants from Chapai Nababganj and Satkhira were adult men, largely stemming from the male-based labour markets of these destinations (i.e. construction sector, rickshaw pulling etc.). On the other hand, both adult men and women migrated from Munshiganj because women had access to network with the job markets in the garment industry and other manufacturing sectors. Given the pre-existing social network with the male job market, it is natural that those households of Chapai Nawabganj and Satkhira would be able to participate in the labour market that have young male members, while families with only young females or for that matter elderly male members will remain largely excluded from migration experiences.

This leads us to conclude that while migrants often perceive the influence of climate change on their migration decisions, it is often difficult for them to distinguish the role of climate change vis-à-vis other social, economic and political factors. Under similar climatic stresses, some link their decisions to migrate with climate change, while others clearly negate any relationship of climate change events with their migration. Migration decisions are also gender, age and culture specific.

Comparison of economic situation of migrant and non-migrant households

Table 7.4 makes a comparison of perceived household financial situation of migrants and non-migrant families. As mentioned earlier, both the migrant and non-migrant families interviewed in this study had moved residence at least once previously. As well, both groups were selected from similar socio-economic backgrounds. However, the table shows that when comparing the responses made from their former and final dwellings, migrants' families were much more likely to report an improvement in their perceptions of their economic situations. Before

Table 7.4 Household financial situation in first and present dwellings

Presence and absence of migration experience and impacts on household financial situation		Always sufficient	Just sufficient	Often insufficient	Total percentage
Migrant	First dwelling	6.1%	45.6%	48.3%	100.0%
	Present dwelling	10.9%	59.6%	29.5%	100.0%
Non-migrant	First dwelling	4.9%	52.1%	43.0%	100.0%
	Present dwelling	2.5%	46.6%	50.9%	100.0%

migration, 6.1 per cent of the migrants reported that their economic status was 'always sufficient'. After migration, this number increased to almost 11 per cent. On the other hand, when interviewed in their former residences, 5 per cent of the non-migrants reported their economic status as being within the 'always sufficient category'. After moving residences, however, only 2.5 per cent of non-migrant families described their economic status as 'always sufficient'. The number of persons in the category of just sufficient also increased for migrants and decreased for non-migrants. This leads us to argue that if households are able to ensure viable relocation in the face of hazards and when livelihood migration of a few members of households has the potential to provide better income and social protection to the migrant households compared to the non-migrants, then migration is one of the many tools of adaptation. Table 7.4 also shows that a section of the households which were trying to adapt locally may trap themselves into occasional or chronic poverty.

Based on these findings, we can argue that livelihood diversification at the local level by a few members of the family along with migration one or a few other members produce better adaptation experience.

Migration in climate change and development policies

Empirical findings presented in different sections demonstrate that in the context of climate change, government, development partners and NGOs are implementing various programmes of planned adaptation. Villages and communities have also undertaken different activities to adapt with the changes that have taken place in their lives and livelihood over the past decade.

Programmes of planned adaptation never addressed the migration issue, whereas autonomous adaptation practices of people incorporated all types of migration as integral parts of their adaptation experiences. We have also seen that households which incorporated migration of one or more household members as an adaptation strategy tended to find themselves in an improved economic situation, and were much more likely to describe their economic situation as 'always sufficient' than households who did not incorporate migration. Based on this finding, in the following section, we will analyse different policies of government and suggest ways to incorporate migration in them. Four types of policies have been reviewed to understand how migration has been perceived. These are (i) policies on climate change, (ii) development policies, (iii) disaster management policies and (iv) migration policies.

Major policies that dealt with climate change and migration are: National Adaptation Programme of Action (NAPA), the Bangladesh Climate Change Strategy and Action Plan (BCCSAP) and National Strategy on the Management of Disaster and Climate-Induced Internal Displacement (NSMDCIID). Initially, the NAPA attempted to reduce the scope of migration, and thus halt undesired 'social consequences' of migration to cities. The updated NAPA document of 2009, however, has omitted the negative references to migration. The BCCSAP of 2008 treated migration of climate change–affected people as a problem of unplanned urbanization. The revised BCCSAP of 2009 highlights the need to understand the dynamics of climate change–induced migration. It suggested resettlement of 'environmental refugees' in the developed countries. Such statements are made from the perspective of presenting migration as a threat to developed countries, some of whom are the major contributors to climate change. The agenda of the policymakers may have been to use this at the negotiating table. However, such a position presents the victims as threat.

In 2015, the Ministry of Disaster Management and Relief of the Government of the People's Republic of Bangladesh prepared the NSMDCIID. This strategy does not cover all aspects of migration. It concentrates on internally displaced populations caused by climatic hazards. It is a comprehensive strategy covering all three phases of displacements: (i) pre-displacement, (ii) displacement and (iii) post-displacement. It commits the rights of displaced to be resettled in the same locality where it is possible; if not, into planned resettlement areas as well as equipping those who choose to migrate on their own nationally or internationally. This can be seen as a new generation document of government that deals with migration of climate-induced displaced population from a rights perspective.

Development policies include the Seventh Five Year Plan (FY2016–FY2020) and the Outline Perspective Plan of Bangladesh 2010–2021 (Vision 2021). These documents highlight the importance of short-term labour migration in the context of the economic development of Bangladesh, and a few of them stressed on increasing the potential of this form of migration (Martin et al., 2013). Most of these documents did not appreciate the contribution of internal migrants. The Government of Bangladesh has yet to develop any comprehensive policy to deal with internal migration in general.

The Seventh Five Year Plan (FY2016–FY2020) of the Government of Bangladesh (Planning Commission, 2015) particularly focused on the need for integrating international migration into development strategy. It suggested measures that are required so that the number of migrants working abroad should be increased from the districts that are lagging behind in sending migrants and receive very little share of foreign remittances. It also highlights the setting up of technical and vocational training institutions in those districts so as to create opportunities for the people to acquire skills and increase their employability in the formal sector jobs in the urban areas of Bangladesh. However, there is hardly any understanding of how migrants and their families can benefit from different development strategies of the Seventh Five Year Plan. This document has a separate section on climate change, but it is yet to explore the link between migration and climate change.

The MOWR (2005) Coastal Zone Development Policy (2005) did not even mention migration, while the Overseas Employment Policy (2006) and the Overseas Employment Act (2013) did not deal with the climate change issues.

Assessments of development partners' interventions in climate change adaptation also show that almost all bilateral and multilateral agencies are deeply committed to climate change issues. They are involved in policy and infrastructure development as well as community-level adaptation programmes. Some of the programmes they have undertaken are oblivious to migration and some others have conceptualized their interventions from the same framework that is targeted to help people to stay in their places of origin (Siddiqui et al., 2013).

We can sum up our policy analysis by stating that except the NSMD-CIID strategy document of MoDMR (2015) for the rehabilitation of the displaced, no other document has yet reflected the potential of labour migration in adapting to different stresses of climate change. The Seventh Five Year Plan (FY2016–FY2020) of course stated that access to international migration increases the opportunity of economic advancement of households who participate in it. The plan also suggests creating access of areas which have lagged behind in international migration. Nonetheless, climate change–affected areas remain largely absent from such discourse.

Conclusions and policy consideration

This paper highlights that all three areas under the study are facing all kinds of environmental and climate change–related hazards. Out of 1,502 migrant and non-migrant households, 75 per cent experienced these stresses in different periods of their life. Half of them faced multiple stresses. In all three areas, people are establishing new practices to adapt with the changes associated with environmental volatility and climate change. These practices combined both planned and autonomous interventions. Satkhira experienced the highest number of planned adaptation interventions, followed by Chapai Nawabganj. In Satkhira, there is a substantial presence of both government and NGO organizations. In Chapai Nawabganj, however, the majority of the programmes are managed by different government ministries and agencies. Experience of planned adaptation is the lowest in Munshiganj, and again they mostly represent government infrastructural programmes. Microcredit NGOs are operational in Munshiganj in a major way, but climate change–related NGOs are largely absent.

All three areas have experiences of autonomous adaptation initiatives. These range from small-scale community infrastructure to changes in working hours and lifestyle. However, in respect to autonomous adaptation practices, the most prominent is the attempts taken for embracing new livelihoods. At the local level in all three areas, non-farm occupations, trade and services have taken over agriculture-based employments. Government adaptation programmes have been successful at developing infrastructure and increasing access to solar electricity or drinking water, but much less so when it came to facilitating the creation of alternative livelihoods.

Among the autonomous adaptation practices, migration of affected households is the most important. We have witnessed examples of relocation of entire and partial villages or individual homesteads into neighbouring or distant villages. These examples allow us to demonstrate that the most effective adaptation in case of loss of homestead is relocation. Such relocations are nothing but examples of migration being an adaptation tool. But not all the attempts of relocation may increase adaptation capacity; some of them are just survival measures. On average, migrant and non-migrant households have shifted their homesteads almost three times. The highest number of homestead relocation was experienced by the people of Munshiganj (3.5 times). A section of relatively well-off households of all three areas have experienced permanent migration. It was most clearly visible in case of Satkhira, where 25 per cent of the affluent Muslim families have migrated to the nearby cities. Better-off Hindu families have migrated to the West Bengal State of India through their network of previous migration of family members. In sum, then not all migration experiences can be termed as adaptation but some of them definitely increase capacity to adapt.

Another major trend is seasonal labour migration. Members of poorer households participate in such migration. Along with sowing and harvesting of paddy, a substantial number are also seasonally migrating to work in brick kilns. There has been an increase in the number of people migrating and the length of their stay has also prolonged. Temporary internal rural-to-urban migration flows are practised in all three areas, with Chapai Nawabganj being the most predominant site of this kind of movement. From certain villages of Chapai Nawabganj, as many as 80 per cent of the working-age male population now cyclically moves between their rural origin and urban destinations. Being close to Dhaka city, both male and female migrants of Munshiganj could ensure work in the formal sector, whereas the majority of the internal migrants of other two origin areas work in the informal sector. Short-term international contract migration, on the other hand, has mostly been experienced in Munshiganj which is less sensitive to climate change. This discussion allows us to conclude that internal displacement, rural-to-rural seasonal and rural-to-urban cyclical migrations are the three forms of movement which are most sensitive to climate change.

It is understood that climate change is taking place and migration is also increasing. Nonetheless, it is almost impossible to distinguish how much of the current migration flows is linked directly to climate change and how much of it is influenced by other reasons. One may argue that along with economic, social, demographic and political factors, environmental factors also influence migration decisions. These influencing factors interact among them and also with micro- and meso-level factors and produce migration decisions. Where the combination of factors is favourable, individuals or communities move. However, conditions in many areas are not as conducive to migration, and thus much less people are able to migrate. An interesting finding of this paper is that those households who have adapted better were those which mixed different local level adaptations with the migration of a few members of the concerned households. Before migration, 6 per cent of the migrant families perceived themselves as always food sufficient, but after migration the number of households who now consider are food sufficient increased to almost 11 per cent. Whereas 5 per cent of the non-migrant families earlier considered themselves to food sufficient category, after moving residences, their percentage has decreased to 2.5. A section of the households which did not take part in migration appear to be trapped in occasional and chronic poverty.

This paper finds a mismatch among the potential of migration as an adaptation tool and government policies of climate change, development and migration. A large number of people are using migration as one of the adaptation tools, but government policy often either discourages migration or remains oblivious to it. Recently, some changes in the government attitude have been observed. The NSMDCIID strategy document of MoDMR has strongly integrated migration into its pre-, during and post-disaster rehabilitation programmes. The Seventh Five Year Plan (FY2016–FY2020) has taken a more positive approach on the use of migration as one of the tools for national development. In the climate change–related policies and strategies such as NAPA and BCCSAP, analysis surrounding migration is far from sufficient. This study, therefore, argues for bringing in changes in climate change policies and strategies. It suggests a delinking of rural-to-urban migration with crime, as well as restraint from framing migration as a threat while negotiating with the global community. Migration needs to be transformed into one of the many planned adaptation tools in development and climate change programmes. Sustainable livelihoods need to be created at the local level and market-oriented skills development training centres should be established in the close vicinity of climate-affected areas. Climate Trust Funds should be allocated on such training. Institutions that process international migrants along with branches of Migrants' Welfare Bank should also be established in those areas.

Urban development policies should have allocation of space for service providers. A policy needs to be framed on internal migration. The current trend of the development of the mega city has to be replaced by the development of multiple growth centres all over Bangladesh. Enhanced connectivity between the growth centres and the hinterland would result in greater mobility of both male and female workforce, and would subsequently cause greater remittance flow. Decentralization would be the key to such policy change.

Notes

1 FGD meeting at Padmapukur Union of Shyamnagar, Satkhira. Conducted in July 2012.
2 Interviewed in July 2012.
3 Interviewed in November 2013.
4 Interviewed in November 2013.
5 Interviewed in July 2012.
6 Interviewed in August 2012.
7 PSF is a simple technology in which water is pumped from a pond and passed through a number of chambers containing sand and gravel. The treated water is usually safe for drinking.
8 Interviewed in August 2012.
9 Interviewed in November 2013.
10 Interviewed in November 2013.
11 Interviewed in November 2012.
12 FGD meeting at Padmapukur Union of Shyamnagar, Satkhira, and Char Paka union of Shibganj, Chapai Nawabganj. Conducted in July–August 2012.

13 Interviewed in November 2013.
14 Interviewed in August 2012.
15 For more information, please see http://en.banglapedia.org/index.php?title= Fodder_Plant
16 Secondary data of 2001 and 2011 Bangladesh census was used in assessing population trends of the *upazilas* affected by climatic stresses such as inland flooding, storm surges and river erosion to extrapolate figures for making long-term predictions for migration and displacements (Kniveton et al. (2013)).
17 Name: Aslam; Age: 30; Sex: Male; Village: Nagerhati, Union Chitrokot; District: Munshiganj. Interviewed in July 2012.
18 Interviewed in July 2012.
19 Interviewed in November 2013.
20 Interviewed in November 2013.
21 Interviewed in August 2012.
22 Interviewed in July 2012.

References

Afifi, T. (2011), 'Economic or environmental migration? The push factors in Niger', *International Migration*, 49(s1), e95–e124.

Ahmed, S. R. (2000), *Forlorn Migrants: An International Legal Regime for Undocumented Migrant Workers*. Dhaka: UPL.

Barnett, J., and Webber, M. (2010), "Accommodating migration to promote adaptation to climate change", World Bank Policy Research Working Paper Series, 5270.

Black, R., Banerjee, S., and Kniveton, D. (2012), "Migration as an Effective Mode of Adaptation to Climate Change", Policy paper for the European Commission. Sussex Centre for Migration Research, University of Sussex.

Ezra, M. (2001), 'Demographic responses to environmental stress in the drought-and famine-prone areas of northern Ethiopia', *International Journal of Population Geography*, 7(4), 259–279.

Hunter, L. M., Luna, J. K., and Norton, R. M. (2015), 'Environmental dimensions of migration', *Annual Review of Sociology*, 41, 377–397.

Kniveton, D., Rowhani P., and Martin, M. (2013), "Climate Scenarios for Bangladesh", Working paper 3, an output of research on climate change related migration in Bangladesh; RMMRU and SCMR, 2013.

Kothari, U. (2003), 'Staying put and staying poor?', *Journal of International Development*, 15(5), 645–657.

Martin, M., Kang, Y. C., Billah, M., Siddiqui, T., Black, R., and Kniveton, D. (2013), "Policy analysis: Climate change and migration in Bangladesh", Working paper 4, an output of research on climate change related migration in Bangladesh; RMMRU and SCMR, 2013.

Martin, M., Siddiqui, T., and Islam, M. T. (2013), *Migration in Bangladesh and Its Sensitivity to Climate Change and Variability*. Sussex: UK & RMMRU: Dhaka. http://migratingoutofpoverty.dfid.gov.uk/files/file.php?name=pb-2-mig.pdf&site-354

McLeman, R., and Smit, B. (2006), 'Migration as an adaptation to climate change', *Climatic Change*, 76(1–2), 31–53.

MEWOE (2006), "Overseas Employment Policy 2006", Ministry of Expatriate Welfare and Overseas Employment, Dhaka, Government of the People's Republic of Bangladesh.

MEWOE (2013), "Overseas Employment Act 2013", Ministry of Expatriate Welfare and Overseas Employment, Dhaka, Government of the People's Republic of Bangladesh.

MoDMR (2015), "National Strategy on the Management of Disaster and Climate Induced Internal Displacement (NSMDCIID)", Prepared by Siddiqui, T., Islam, M. T., and Akhter, Z. (2015). Dhaka: CDMP-II, Ministry of Disaster Management and Relief.

MOEF (2006), "National Adaptation Programme of Action, Updated Version of 2005", Dhaka, Ministry of Environment and Forests, Government of the People's Republic of Bangladesh.

MOEF (2008), "Bangladesh Climate Change Strategy and Action Plan, 2008", Ministry of Environment and Forests, Dhaka, Government of the People's Republic of Bangladesh.

MoWR (2005), "Coastal Zone Policy 2005, Ministry of Water Resources", Dhaka, Government of the People's Republic of Bangladesh.

Planning Commission (2015), "The Seventh Five Year Plan (FY2016–FY2020) Accelerating Growth, Empowering Citizens", Government of the People's Republic of Bangladesh.

Siddiqui, T., Bhuiyan, R. A., Black, R., and Kniveton, D. (2013), "Migration-from Threat of Climate Change to Climate Change to Adaptation Tool: Policy Choices for Development Partners of Bangladesh", Policy Briefing paper No. 6, an output of research on climate change related migration in Bangladesh; RMMRU and SCMR, 2013.

Siddiqui, T., and Billah, M. (2014), "Adaptation to climate change in Bangladesh: Migration the missing link", in S. Vachani and J. Usmani (eds.), *Adaptation to Climate Change in Asia* (pp. 117–141), Cheltenham, UK: Edward Elgar Pub.

Sidddiqui, T., and Mahmood, R. A. (2015), "Impact of Migration on Poverty and Local Development in Bangladesh", SDC and RMMRU, Dhaka.

Tacoli, C. (2009), 'Crisis or adaptation? Migration and climate change in a context of high mobility', *Environment and Urbanization*, 21(2), 513–525.

Warner, K., and Afifi, T. (2014), 'Where the rain falls: Evidence from 8 countries on how vulnerable households use migration to manage the risk of rainfall variability and food insecurity', *Climate and Development*, 6(1), 1–17. doi: 10.1080/17565529.2013.835707

World Bank (2010), *The Economics of Adaptation to Climate Change, Environment Department*. Washington, DC: World Bank.

World Bank (2011), *The Cost of Adapting to Extreme Weather Events in a Changing Climate*, Bangladesh Development Series Paper No. 28. Washington, DC: World Bank.

8 Migration in response to environmental change

A risk perception study from Sundarban Biosphere Reserve

Avijit Mistri and Bhaswati Das

The estimates of climate migrants vary but the figures are daunting as researchers predict the number in hundreds of millions by middle of the 21st century (Myers, 2005; UNDP, 2007; Stern, 2006). A large population in India, Bangladesh, China and Vietnam are living in low-lying coastal areas and islands. One-quarter of Bangladesh's population (around 35 million) lives in the coastal floodplain (Stern, 2006). China's 41 per cent population, 60 per cent wealth and 70 per cent of its megacities are located in coastal areas (UNDP, 2007). In the fag end of 2000s, IPCC (2007) in its fourth Assessment Report by Group II warned that Asian mega deltas especially the Ganges-Brahmaputra delta in Bangladesh and West Bengal in India is in the most vulnerable condition due to climate change, sea level rise and more intensification of coastal storms. This vulnerability would be increased with low human adaptive capacity and high level of exposure. The present study area, the Indian Sundarban is part of the Ganges-Brahmaputra delta comprised of 102 islands. It is the largest halophytic mangrove forest in the world. It contains magnificent biodiversity and outstanding universal value, and is also famous for in situ conservation of some globally endangered species. It is part of the World Network of Biosphere Reserves under UNESCO's Man and the Biosphere Programme. A total of 9,600 sq. km, including 4,200 sq. km forested area and 5,400 sq. km inhabited area, is recognised as Sundarban Biosphere Reserve (2001). The region is the habitat to 4.4 million people.

Geographically, Sundarban is an archipelago (54 inhabited and 48 forested islands) along the Bay of Bengal coast. Most of the inhabited islands are juvenile and unstable in nature. Before completion of the build-up process through siltation, people erected mud embankments and occupied the land on the bank of the rivers and settled down. The absolute height of the islands varies from 0.90 to 2.11 m from the mean sea level (MSL). The islands, therefore, are highly susceptible to the global warming and hydro-climatic events. The region has witnessed the recent occurrence of Severe Cyclonic Storms (SCS) – Aila in 2009 and Sird in 2007 (Super Cyclonic Storm), Bijli in 2009 (SCS) and Phailin in 2013 (extremely SCS). In addition to this, the inhabitants of Sundarban depend on natural resource–based activities which are highly susceptible to the climatic stimuli. Nearly 59 per cent of the workers are engaged in farming and 5 per cent are involved in fishing (Census, 2011) here.

Outmigration is a prominent livelihood strategy in Sundarban. At least one member from three-fourth of the households (HHs) migrate out from the area in search of work (Mistri, 2013, p. 94; Ghosh, 2012, p. 77). They prefer interstate migration over inter-district and intra-district migration (Mistri, 2013). The lack of livelihood opportunities is the key driver of migration from Sundarban. Agriculture is the primary means of livelihood in Sundarban. Rice is the principal crop. Its yield is 2,204 kg per hectare (three years moving average 2008–2009 to 2010–2011) which is lower than the state average that is 2,573 kg per ha. The yield has gradually been declining at the rate of 0.15 per cent per year from 1990–2000 to 2010–2011 in South 24 Parganas (B. A. E. & S., 2011). Between 2009 and 2013, there was a complete crop failure due to high salinisation posed by Aila. One storm surge causes at least two to three years of complete crop failure and subsequently more than two years of low yielding rate. Besides, most part of Sundarban farming practises rain-fed single cropping during monsoon. The farmers thus experience long lean periods in a year. Another principal means of living in Sundarban is fishing that has been hampered by the environmental legislations (Mistri and Das, 2015). Fishing in Sundarban includes catching fish, prawn, crab and other aquatic animals and collecting non-timber forest products (NTFPs), such as honey, bee wax, firewood, tannin bark, *nypa fruticans* (golpata) etc. While introducing new rules and regulations for better conservation of the reserved forest, implementation has been made stringent affecting fishing severely. The environmental legislations in Sundarban have prioritised environmental sustainability ignoring the effect on means of livelihood. There is also a three-month lean period in fishing from the month of April to July as it is the breeding period of the fish. Along with stringent environmental legislations, hydro-climatic hazards, degrading riverine environment and human-tiger conflict have made the life of the population more hazardous. People of Sundarban hardly get work for more than six months in a year. They are forced to diversify their jobs, but there is a lack of job opportunities in that region. People of Sundarban mitigate the fragile and unsustainable economic base by migrating out.

In Sundarban, the influence of environment on outmigration cannot be denied. Environmental migration study still lacks methodological base to identify the influence of environmental change on migration. People are found to be immobile in extreme weather events due to the low level of asset possession (Black, et al., 2012). Migration is a multi-causal phenomenon. The framework associated with exploring the probable drivers of migration and then examining how environmental change influences them is a more convincing way of study. The Sustainable Livelihood Approach (SLA) by DFID (2001) is widely popular in this context. However, to examine the impact of environment on the drivers or livelihood issues of migration, or a nexus between them is widely multi-pronged and it involves methodology of both natural and social sciences. For example, as part of natural science it tries to explore to what extent one unit change in temperature or rainfall affects the yield of rice or wheat, or to what extent one ppt pH decrease affects the shrimp or prawn production. However, they are beyond the scope of the study of the present paper. The method involved in the present task examines the association between the risk perception of environmental change and livelihood issues. The risk perception is the judgement of people when they are asked to characterise and evaluate hazardous activities (Slovic and Peters, 2006; Slovic, 1987). It puts emphasis on people's cognition which guides the judgement or decision making regarding stimuli, or influences of stimuli on a livelihood choice. In the study, people's risk perception on livelihood exposes the linkage between environmental change and livelihood miti gation through migration from Sundarban. The livelihood is comprised of assets, means of living (activity) and capability (Camber and Conway, 1991). Among the three facets, activities of the people play the most significant role for migration which is the focus of the present study.

Objective

The study emphasises the subjective appraisal of the environmental change in Sundarban and its interlink with outmigration. Whether the inhabitants of Sundarban is concerned about the environmental change in Sundarban or not is the first question of the study. The study focuses on people's overall degree of concern regarding the environmental change in Sundarban. Therefore, people's perceived risk in farming and fishing in the changing environment is brought under the purview of the discussion. The association between environmental risk perception and the decision of migration are examined side by side.

Moreover, the perceived risk of the migrants is compared with a controlled group that is non-migrants.

Methodology and data source

The common approach for studying risk perception is the psychometric paradigm that includes psycho-physical scaling, affective imagery, multivariate analysis and so on. The techniques of psycho-physical scaling, such as environmental risk perception indices and the area of concern, are incorporated into the study. Simple statistics like mean score, standard error of mean (*SEM*) and one sample *t*-test are used.

The study is based on the field investigation conducted through structured questionnaire schedule. Census villages were considered as the lowest geographical units for sampling and a respondent in a household (HH) within a census village was considered as the lowest unit of information. A total of 400 respondents (from 400 HHs) from eight census villages (50 respondents from each) distributed over four selected administrative blocks were surveyed. A total of 19 blocks –13 from the district of South 24 Parganas and 6 from the district of North 24 Parganas – comprise the inhabited part of the Sundarban Biosphere Reserve (SBR). Only the blocks under South 24 Parganas were taken into consideration because the Kolkata urban agglomeration highly influences the blocks under the North 24 Parganas and these are socio-economically different from the rest of the Sundarban (see Map 8.1). The selected census villages are Lahripur (1) and Sonagar (2) from Gosaba block, Parbatipur (3) and Joygopalpur (4) from Basanti block, Radhakantapur Abad (5) and Pashchim Jatardeul (6) from Mathurapur II and Bijaybati (7) and Amrabati (8) from Namkhana block.

Total samples are divided into two equal groups – namely, control group and exposed group. The control group is that group of households where no member had migration experience during the last 365 days before the date of survey. The exposed group is that group where at least one member of the household experienced migration before the date of survey. A total of 400 respondents, 200 from each group, were selected. All the respondents were from the working age, 15 to 59 years. In the case of the control group, the head of the household or any working person in the absence of the head of the household was considered as the respondent. In the exposed group, only the migrating person who was present at the time of the survey and bearing the migration status till date – this means he or she came home to an occasion or a purpose not to return permanently – was considered as a respondent. If more than one was present, the elderly migrant was selected.

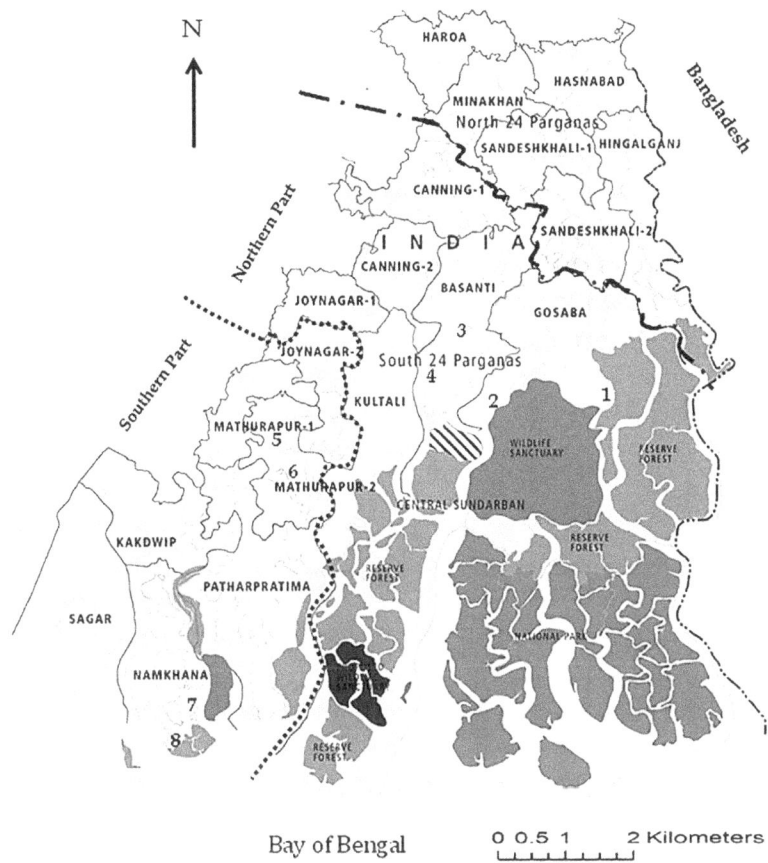

Map 8.1 Schematic plan of sampling, Sundarban Biosphere Reserve
Source: Ghosh (2012), p. 8

Results and discussion

Risk perception index

The environmental concern of the people of Sundarban is attributed to four categories (Table 8.1) such as holistic concern, the concern for local impacts, the concern for non-human nature and the seriousness of current impact around the world. These attributes help to understand how people's concern varies from geographically distant places and

Table 8.1 Environmental change Risk Perception Index (RPI) in Sundarban

Attribute	Indicators	Mean ($\overline{x}$)	std. error of mean (SEM)	Test value: $\mu = 3$		
				Mean diff. ($\overline{x} - \mu$)	t	p (2-tailed)
Holistic concern	How concerned are you about the environmental change in Sundarban?	2.64	0.062	−0.357	−5.793	0.000
Local impacts	Do you think that the following incidences will occur due to the environmental change during the next 25 years?					
	Do you think your means of living (farming /fishing / others) will be hindered?	2.85	0.065	−0.152	−2.347	0.019
	Water shortage will occur where you live?	2.84	0.068	−0.160	−2.360	0.019
	Your standard of living will decline?	2.67	0.052	−0.335	−6.401	0.000
	Your chance of getting a serious disease will increase?	3.09	0.060	0.087	1.469	0.143
Non-human nature	Do you think the environmental change-					
	Destroy mangrove forest?	2.35	0.060	−0.648	−10.861	0.000
	Affect the wild animals (tiger, monkey, deer, crocodile, turtle and tortoise, birds, and others)?	2.50	0.059	−0.505	−8.525	0.000
Worldwide impact	How serious are you about the current impact of environmental change in Sundarban on worldwide?	2.55	0.066	−0.453	−6.876	0.000
	RPI	**2.68**	**0.041**	**−0.315**	**−7.645**	**0.000**

Scale range from 1 (to no extent) to 5 (to a very great extent); $n = 400$, $df = 399$, $p<0.05$ – Statistically significant

Source: Computed from Field Survey Data; based on Leiserowitz (2006)

peoples that is in effect on the outside of the Sundarban or the world-wide impact or for non-human nature like mangrove forest and wild animals. The cognitive judgement of 400 respondents was collected by different sets of questions structured with a five-point Likert-type scale (Table 8.1). Each respondent was requested to respond with any one of the options, namely 'to a very great extent', 'to a great extent', 'to some extent', 'to a little extent' and 'to no extent'. The values assigned to these options are 5, 4, 3, 2 and 1, respectively. The mean is 3.0 [(5+4+3+2+1) /5 classes]. The score 3.0 is deemed to be a normal level of concern. The variable with a (sample) mean score ($\bar{x}$) less than 3.0 is considered as the low risk perception or low impact of that particular stimulus. Mean score above 3.0 is considered as the great risk perception. The one sample *t*-test is used to determine whether the sample comes from a normal population, which has a mean (μ) of 3.0. That means the test examines the significant difference ($p < 0.05$) between the sample mean ($\bar{x}$) and hypothesised population mean ($\mu = 3.0$). In addition to that, the mean difference ($\bar{x}-\mu$) expresses how much lower or higher the $\bar{x}$ is compared to μ (3.0) at 95 per cent confidence interval (*CI*). The SEM infers how precisely the sample predicts the true mean of the population (not to the hypothesised mean) or how close the sample is to the true population mean.

The environmental change risk perception (index) score of the respondent is 2.68 (Table 8.1). The one sample *t*-test (–7.645) reveals that the RPI score is significantly different ($p = 0.000$) from the hypothesised

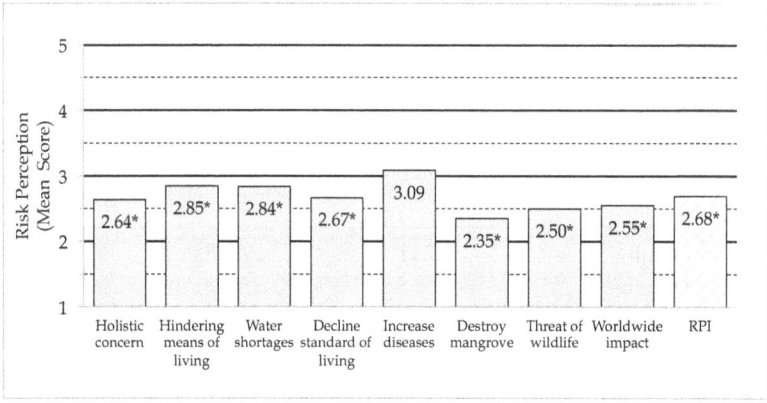

Figure 8.1 Islander's risk perception in the current impact of climatic conditions

Note: * Statistically significant ($p < 0.05$) difference from $\mu = 3.0$

Source: Computed from Field Survey Data

population mean (μ = 3.0). It is significantly lower by 0.315 at 95 per cent *CI* (–0.40 to –0.23) level of confidence. The respondents, therefore, perceive environmental change in Sundarban as a lower risk. The mean scores of the seven indicators out of eight are significantly lower than the hypothesised mean (μ = 3.0). Only one indicator, that is, chances of getting a serious disease is not statistically significant difference – that means it is equal to the μ = 3.0 or average level of concern. Though the mean scores of the seven indicators are significantly lower than the hypothesised mean, these are not widely varying from the same. The mean differences ($\bar{x} - \mu$) are not lower than 0.50 and implies that people's environmental concern is not alarmingly low.

Moreover, the highest mean score with $p<0.5$ among the indicators infers that islanders are most concerned regarding the hindrance in their means of livelihood ($\bar{x}$ = 2.85) (Figure 8.1). The next two important concerns are water shortage ($\bar{x}$ = 2.84) and the declining standard of living ($\bar{x}$ = 2.67). They are relatively less concerned about the threats faced by non-human nature like mangrove forest ($\bar{x}$ = 2.35) and wild animals ($\bar{x}$ = 2.50) such as tiger, monkey, deer, crocodile, turtle and tortoise, birds. They are little concerned about the worldwide impact ($\bar{x}$ = 2.55) of the environmental change on Sundarban.

Areas of concern

The mentioned conclusion is supported by the result of a separate question that was asked to the respondents to specify which area of concern put them into greatest anxiety due to environmental change in Sundarban (Table 8.2). A clear majority of respondents (67 per cent)

Table 8.2 Areas of concern

Most concerned about the impact on?	Frequency	Percent	Cumulative percentage
You and family	251	62.8	62.8
Local community	15	3.8	66.5
Sundarban Biosphere Reserve as a whole	60	15.0	81.5
People all over the world	40	10.0	91.5
Non-human nature	9	2.3	93.8
Not at all concerned	25	6.3	100.0
Total (*n*)	400	100	

Source: Field Survey

are most concerned about the impact on themselves, their family and their local community. Total 25 per cent are most concerned about the effects on Sundarban as a whole and on people around the world. Only 2 per cent are concerned about non-human nature.

However, it is important to notice that to understand the big national and environmental issues one should consider the direct local relevancies. Without the local people's recognition, addressing environmental problems remain in vain. Nowadays, in the politics of global warming, most of the nations are more concerned about carbon emission of other countries rather than their own level of the same. Climate change is unlikely to become a high-priority national issue until those countries consider themselves personally at risk (Leiserowitz, 2006).

Environmental change risk perception: migrant vs. non-migrants

The RPI score for migrants ($\bar{x}$ = 2.70, n = 200, SEM = 0.062) and non-migrants ($\bar{x}$ = 2.67, n = 200, SEM = 0.054) is significantly ($p<0.05$) lower than the normal level of concern (μ = 3.0). Perceived risk of migrants is slightly greater than non-migrants (Figure 8.2). The migrants are more concerned about their occupation. The mean scores are 2.92 and 2.78 for migrants and non-migrants, respectively. It may be one of the potential causes for the decision to migrate. The declining standard of living and the incidence of increase in diseases are little more of a concern of non-migrants as they are firmly attached to the origin. Moreover, the migrants have shown more awareness about the environmental impact

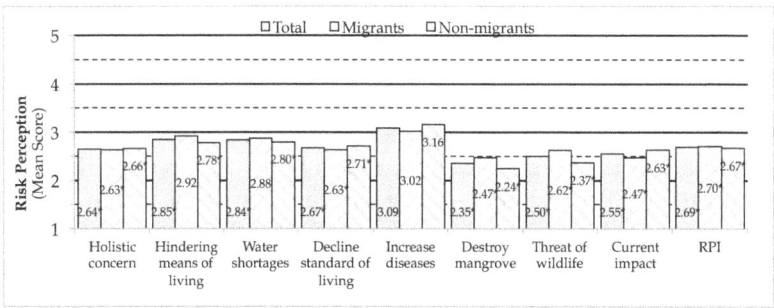

Figure 8.2 Environmental change risk perception between migrants and non-migrants

Note: * Statistically significant ($p<0.05$) difference from μ = 3.0.

Source: Computed from Field Survey Data

on wildlife and forest compared to the non-migrants. The reason may be the migrants' acquaintance regarding the fame of the Sundarban, or being recognised by that fame while working at the destination. However, an independent sample t-test (t = 0.416, df = 398) between the migrants and non-migrants in terms of overall risk perception reveals that there is no statistically significant difference (p = 0.677) in the mean score between the two groups. The effect size, that is, the magnitude of differences between the groups is very low (*Eta Squared* = 0.0004 or 0.04 per cent).

Environmental risk perception in means of living

Farming and fishing is confronted by climatic change in different ways such as deteriorating environment, declining production or collection, degrading quality of production, increasing diseases, pests and other affecting agents, reducing productivity of labourers etc. (Juana, et al., 2013; Farauta, et al., 2011; Falco, et al., 2011). However, people's concern in their respective occupations is the outcome of their continuous observed experiences. In the study, last five years' actual experiences of the respondents were collected by different sets of questions designed with a five-point grading scale.

Risk perception in farming

In the study, the incidences of environmental change in agriculture manifest through uncertainty at the onset of the season, effects of extreme climatic events and climatic processes and increase in farming problems. A total of 17 distinct perceptions of climate change consequences have been collected from the respondents (Table 8.3). The recall period is at least five years. Risk perception is evaluated with a five-point Likert-type scale which ranges from 1 (to no extent) to 5 (to a very great extent).

The index score for risk perception in farming is 2.95 which is significantly different (p = 0.046) from the test value, μ = 3.0. But the score is not far below the average level of concern (μ = 3.0) and the difference is only 0.052. The t value is −1.999, which is just at the margin of the critical value of t-test that ranges from ±1.96 or to round off, ±2 for 95 per cent level of *CI*. Therefore, the score can be treated as normal. The respondents in the study perceive moderate risk in farming in the current climatic condition.

The low value of *SEM* of the indicators gives an indication of the proximity of the sample mean ($\bar{x}$) to the true population mean (not to hypothesised mean), or the high sample accuracy in the study. The mean scores of two variables, namely increasing the risk of cyclone

Table 8.3 Environmental risk perception in farming

Do you perceive any change in the following phenomenon/incidences in Sundarban during last five years?

Sl. no.	Incidences	Mean ($\bar{x}$)	std. error of mean (SEM)	Test value: $\mu = 3$		
				Mean diff. ($\bar{x} - \mu$)	t	p (2-tailed)
A. Uncertainty in the onset of farming season						
1	Delay in onset of monsoon	2.50	0.059	−0.502	−8.576	0.000
2	Unusual early monsoon followed by weeks of dryness	2.75	0.047	−0.255	−5.477	0.000
3	Erratic rainfall in monsoon season	2.75	0.058	−0.255	−4.335	0.000
4	Heavy and long period of rainfall	3.12	0.058	0.115	1.987	0.048
5	Less rainfall in monsoon	2.18	0.062	−0.820	−13.274	0.000
6	Long period of dry season	2.79	0.051	−0.210	−4.131	0.000
	Average (1 to 6)	**2.68**	**0.034**	**−0.321**	**−9.393**	**0.000**
B. Effect of climatic events and processes						
7	High temperature in summer	3.90	0.051	0.898	17.734	0.000
8	Severity in winter	3.40	0.057	0.400	7.008	0.000
9	Long spell of fog/smog	2.74	0.057	−0.263	−4.637	0.000
10	Overflowing of streams/rivers (flood)	2.76	0.067	−0.243	−3.612	0.000
11	Increasing risk of cyclone (number/intensity)	2.95	0.064	−0.050	−0.777	0.438
	Average (7 to 11)	**3.15**	**0.038**	**0.149**	**3.927**	**0.000**
C. Increase in farming problem						
12	Declining yield rate, Rice (kg/bigha)	2.86	0.058	−0.145	−2.512	0.012
13	High rate of pest incidence in farming	3.29	0.061	0.290	4.793	0.000
14	Increasing diseases in farming	2.97	0.067	−0.030	−0.449	0.654
15	Rotten of rice (Aman) sapling in monsoon	2.65	0.059	−0.350	−5.946	0.000
16	Loss of soil fertility	3.20	0.053	0.203	3.811	0.000
17	Increasing soil salinity	3.33	0.060	0.325	5.383	0.000
	Average (12 to 17)	**3.05**	**0.038**	**0.049**	**1.292**	**0.197**
	Index (Average 1 to 17)	**2.95**	**0.026**	**−0.052**	**−1.999**	**0.046**

Scale range from 1 (to no extent) to 5 (to a very great extent); $n = 400$, $df = 399$, $p < 0.05$ – Statistically significant

Source: Computed from the Field Survey Data

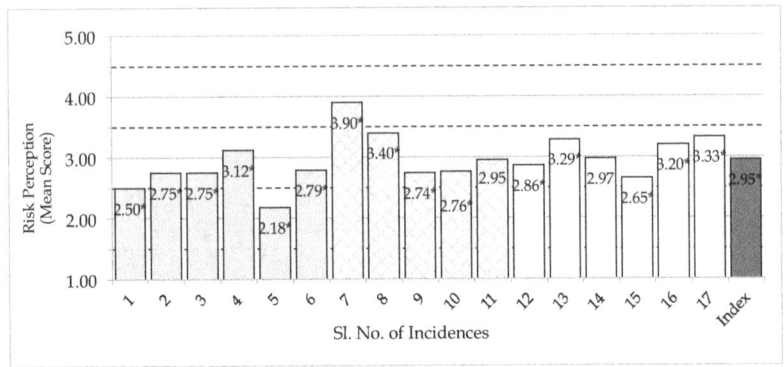

Figure 8.3 Environmental risk perception in farming

Note: * Statistically significant (*p*<0.05) difference from *μ* = 3.0.

Source: Computed from Field Survey Data

($\bar{x}$ = 2.95) and diseases related to farming ($\bar{x}$ = 2.97), are not small enough to make a statistically significant difference. Thus, these are treated as a normal level of risk perception. Apart from these two variables, all are significantly different from the test value. The six indicators among them, such as heavy and long period of rainfall ($\bar{x}$ = 3.12), high temperature in summer ($\bar{x}$ = 3.90), severity in winter ($\bar{x}$ = 3.40), high rate of pest incidence in farming ($\bar{x}$ = 3.29), loss of soil fertility ($\bar{x}$ = 3.20) and increase of soil salinity ($\bar{x}$ = 3.33) are significantly greater than the hypothesised mean (μ = 3.0). The rest are shown in Figure 8.3.

Heavy and long period of rainfall in monsoon is a great concern for the initiation of sowing. This finding is supported by Hazra, et al., (2010, p. 54) who noted that monsoon rain has significantly increased in Sundarban during 2001–2008 at the rate of 0.0041 mm per hour. Heavy rainfall with a long spell at the beginning of farming hampers the preparation of wet bed and sowing of rice; even after transplanting, the sapling is rotting. The very poor sewage system in Sundarban worsens the effect. In addition to that, erratic rainfall in the monsoon and long dry season at times in the year affect the onset of cultivation to some extent.

The sea surface temperature (SST) in Sundarban has been increasing at the rate of 0.4°C to 0.5°C per decade (Hazra, et al., 2010; Mitra, et al., 2009). Though the pace is very slow, it is higher than the global average that is 0.06°C per decade; even greater than the Indian Ocean, which is 0.2°C per decade (IPCC, 2007). Besides, the air temperature has been increasing at a more alarming rate that is 0.1058°C per year

(1°C per decade) during 2002–2009 (Hazra et al., 2010, p. 51). This concern glares in the cognitive minds of the respondents. The temperature increase in summer and its severity in winter influence farming in Sundarban to a great extent. Beyond a certain range of temperature, yield gets reduced. The declining yield (kg per ha) of rice, that is –0.15 per cent per year (three years moving average) during 1990–2000 and 2010–2011 in S 24 Pgs (BAE&S, 2010), may be the probable consequence of that. In addition to that, high temperature interferes with the ability of plants to get and use the moisture and accelerates transpiration and soil evaporation. In Sundarban, sources of irrigation (ponds and canals) are limited. Increase in temperature creates extreme water shortages in summer. Therefore, attempting Rabi cultivation is risky if the farmers do not have plenty of water storage. Moreover, the warm-humid summer and cold-dry winter are the salient climatic characteristics of Sundarban. In extreme weather conditions, farmers are easily exhausted that reduces labour productivity. The extreme climatic event, tropical cyclonic storm, immensely affects the farming. The frequency of Severe Cyclonic Storm (SCS) has significantly increased in the Bay of Bengal during the past 129 years (1877–2005) during the intense cyclone months (May, October and November) – November especially accounts for 25–26 per cent rising of SCS per 100 years (Singh, 2007; Singh, et al., 2001). Apart from frequency, the intensity of storms has also increased in the last decade, from 1999 to 2005 (Hazra, et al., 2010).

The increasing soil salinity in Sundarban is one of the great concerns of the farmers. The rating of the incidence ($\bar{x}$ = 3.33) is most likely among the issues related to the farming problem in Sundarban. The Northern Sundarban has witnessed nearly 8.0 ppt of soil salinity and the southern part which is in close proximity to the coast has registered 8.0 to 20.0 ppt (Ghosh, 2012). The salinity beyond 6.0 ppt hampers rice cultivation. In Sundarban, it exceeds the permissible limit. Moreover, higher salinity is one of the leading causes of declining soil fertility ($\bar{x}$ = 3.20) in Sundarban. Apart from soil salinity and fertility, increasing pest incidents ($\bar{x}$ = 3.29) and diseases ($\bar{x}$ = 2.97) are somewhat likely concerns among the farmers. Overall, the effect of weather events and processes ($\bar{x}$ = 3.15, t = 3.927, p = 0.000) are more cause for concern for the farmers compared to the core farming problems ($\bar{x}$ = 3.05, t = 1.292, p = 0.197) and the onset of the monsoon ($\bar{x}$ = 2.68, t = –9.393, p = 0.000).

Risk perception in farming and migration

The mean score of migrant for environmental risk perception in farming is 2.883 which is significantly (t = 3.017, p = 0.003) lower than

Table 8.4 Independent sample *t*-test (risk perception in farming and migration)

Statistics	Index (Risk in farming)	
	Migrants	Non-migrants
Mean	2.883	3.012
SEM	0.039	0.035
n	200	200
Mean Difference (Mig. – Non-mig.)	–0.129	
t (Independent sample)	–2.486	
p (2-tailed)	0.013	
Eta Squared (Effect size)	0.015	
df	398	

Source: Computed from Field Survey Data

the normal level of concern ($\mu = 3$). But, non-migrants ($\bar{x} = 0.301$, $t = 0.356$, $p = 0.722$) perceive moderate risk in farming.

The comparison between migrants and non-migrants reveal that there is a statistically significant difference ($t = -2.486$, $df = 398$, $p = 0.015$) between the groups (Table 8.4). The migrants perceive risk lowered by 0.129 compared to the non-migrants at 95 per cent *CI*. The magnitude of the difference between them (*Eta Squared* = 0.015) is low. Only 1.5 per cent variance in risk perception in farming is explained in the *t*-test. However, comparatively slightly higher perceived risk among the non-migrants can be attributed to intense attachment with origin by the occupation. It is also an indication of potential migration from Sundarban.

Risk perception in fishing

Another important means of livelihood in Sundarban is fishing. Fishing includes both the collection of fish and the collection of non-timber forest products (NTFPs). The fishermen's concern in Sundarban can be attributed to two categories – worsening of the collecting environment and effect on production. A total of 268 respondents (out of 400), those who were directly and indirectly involved with fishing, reported their concern in a five-point grading scale. The recall period of the respondents also remains the same that is at least five years.

The mean score ($\bar{x}$) for risk perception in fishing is 2.96 which is not statistically significantly ($p = 0.246$) different from $\mu = 3.0$ (Table 8.5). The mean difference is ($\bar{x} - \mu$) is –0.040. Therefore, the respondents perceive a moderate level of risk in fishing. Fishermen are highly concerned

Table 8.5 Environmental risk perception in fishing

Do you perceive a change in the following phenomenon/incidences in Sundarban during last five years?

Sl. no.	Incidences	Mean (x̄)	std. error of mean (SEM)	Test value : μ = 3		
				Mean diff. (x̄– μ)	t	p (2-tailed)
A. Worsening of the collecting environment						
1	Increasing risk of cyclone/low pressure/ gusty wind	2.94	0.068	–0.060	–0.872	0.384
2	Increasing risk of flash-flood/high tide	2.84	0.067	–0.160	–2.399	0.017
3	Increasing risk of tidal force/river or ocean current	2.92	0.063	–0.078	–1.237	0.217
4	High humidity and temperature in summer	2.62	0.066	–0.384	–5.789	0.000
5	Severe winter (windy and low temperature)	2.81	0.061	–0.194	–3.171	0.002
6	Declining depth of rivers and creeks	2.92	0.064	–0.078	–1.220	0.224
	Average (1 to 6)	**2.84**	**0.039**	**–0.159**	**–4.053**	**0.000**
B. Effect on production						
7	Declining fish (fish/ prawn/crab) production	3.76	0.084	0.761	9.046	0.000
8	Reduction of fish/prawn/ crab stocks and varieties	3.49	0.088	0.485	5.488	0.000
9	Increasing trash fish	3.34	0.079	0.343	4.327	0.000
10	Faster spoilage of the catch	1.96	0.067	–1.041	–15.554	0.000
11	Increasing salinity of river/ocean water	2.95	0.062	–0.052	–0.842	0.400
12	Decrease the water transparency (or green colour)	2.76	0.062	–0.239	–3.868	0.000
13	Environmental regulations affect the fishing	3.18	0.081	0.183	2.270	0.024
	Average (7 to 13)	**3.06**	**0.041**	**0.063**	**1.521**	**0.130**
	Index (1 to 13)	**2.96**	**0.034**	**–0.040**	**–1.162**	**0.246**

Scale range from 1 (to no extent) to 5 (to a very great extent), $n = 268$, $df = 267$, $p < 0.05$ – Statistically significant

Source: Computed from the Field Survey Data

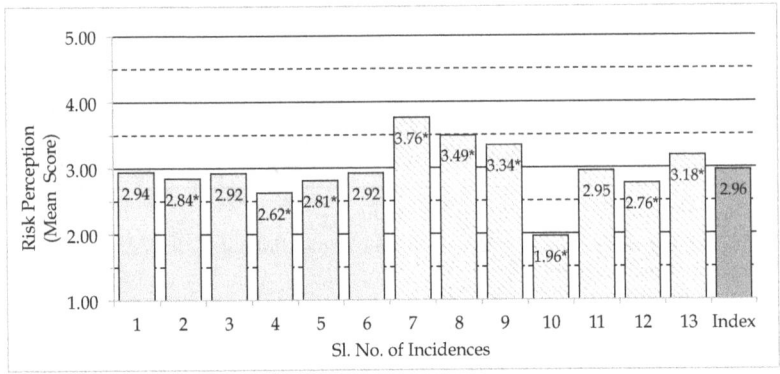

Figure 8.4 Environmental risk perception in fishing

Note: * Statistically significant (*p*<0.05) difference from μ = 3.0.

Source: Computed from Field Survey Data

regarding the declining of fish production ($\bar{x}$ = 3.76, *p* = 0.000) (Figure 8.4). Reduction of fish stocks and its varieties is also a great concern ($\bar{x}$ = 3.49, *p* = 0.000).

Moreover, increasing trash fish[1] ($\bar{x}$ = 3.34, *p* = 0.000) has declined both the quality and quantity of production. Department of Sundarban Affairs (in 2009) reported that during the last 15 years, fishing effort has been doubled but Catch per Unit Effort (CPUE)[2] has drastically declined to 58–65 kg per haul from 150–200 kg per haul. During 1995–2000, tiger shrimp was abundantly available in rivers and creeks and was easily caught by drawing net with the finest weave. Its collection was also economically profitable. One-inch long tiger shrimp was sold at Rs 2–5 per piece in 2000. Near about four lakh people, especially women and children were involved, and approximately 1,500 to 3,000 million shrimps were caught during 1995–2000. The rampant catch had severely destroyed the riverine ecosystem enhancing river bank erosion, destroyed mangroves, killed immature carps and eggs etc. It, therefore, was a short-lived activity, and tiger shrimp was extinct within five years. Moreover, it is a well-known fact that environmental legislations are one of the great barriers of fishing in Sundarban (Mistri and Das, 2015). This phenomenon has resonated in the concerns of fishermen. The mean score for environmental legislations tends to a great extent of risk perception ($\bar{x}$ = 3.18, *p* = 0.024).

Like farming, extreme weather events, especially the increasing risk of cyclonic storms, threaten fishing. The fishermen perceive the moderate extent of risk ($\bar{x}$ = 2.94, p = 0.384) in it. Fishermen often go to catch or collect NTFPs for a week. But in severe cyclone-prone months, such as May, October and November, they often have to leave the fishing ground before reaching the expected schedule. Every year, some fishermen go missing from the ground, victimised by the gusty wind/cyclonic storms. In addition to cyclonic storms, very low temperatures along with windy and foggy winter decline production. Meanwhile, severe warm, humid summer dehydrates and weakens the fishermen, and easily spoils the catch.

Apart from weather, fishing depends on riverine conditions. The decline of river and creek depth ($\bar{x}$ = 2.92, p = 0.224) and an increase in tidal force or current ($\bar{x}$ = 2.92, p = 0.217) are also concerns of the fishermen (moderate extent) in Sundarban. The shallow river channels are not considered as suitable fishing ground. In shallow rivers and creeks, water passes through a particular channel at the time of low tide resulting in increased tidal force. Besides, sand bars are hampering the free movement of fish and seasonal migration of some species. Overall, the respondents are more concerned regarding the issues related to the low production of fish and NTFPs ($\bar{x}$ = 3.06, p = 0.130) compared to the deterioration of the collecting environment ($\bar{x}$ = 2.84, p = 0.000).

Risk perception in fishing and migration

Both migrants ($\bar{x}$ = 2.916, t = −1.810, p = 0.073) and non-migrants ($\bar{x}$ = 3.003, t = 0.068, p = 0.946) perceive more or less equal risk in fishing that is moderate level of risk (μ − 3.0). Therefore, there is no statistically significant difference between the groups (Table 8.6).

Table 8.6 Independent sample t-test (risk perception in fishing and migration)

Statistics	*Index (Risk in fishing)*	
	Migrants	*Non-migrants*
Mean	2.916	3.003
SEM	0.046	0.050
n	132	136
Mean Difference *(Mig. – Non-mig.)*	−0.087	
t (Independent sample)	−1.282	
p (2-tailed)	0.201	
Eta Squared (Effect size)	0.006	
df	266	

Source: Computed from Field Survey Data

Conclusion

Nowadays, climate or environment change occupies the centre of a wide debate; however, there is data limitation (time scales) to prove the rates of change of parameters as evidence. Sundarban is also not an exception. Until now, most of the data sets that have been used in support of climate change in Sundarban is not longer than a climatological normal (30 years). Moreover, most of the environmental indicators have recorded a minimal change and are highly influenced by local factors. There is nothing to worry about the present trend of climate in Sundarban. It may be equilibrated through its dynamic mechanism within the system in the near future. Therefore, it is argued that whatever changes in weather variables have been observed should be defined as climate variability instead of climate change. However, whether climate change or climate variability is observed in Sundarban, people are not free from its ill effects. They perceive environmental risk to affect their livelihood moderately. They are more concerned about micro-risk like their own family, community and local incidents rather than a holistic concern (Sundarban as a whole) about geographically distant places and peoples, such as people around the world, and non-human nature such as wildlife and mangroves. Cyclones, water surges, salinisation, rainfall and temperature changes are more causes of concern affecting their means of livelihood compared to other environmental phenomena.

The principal means of living – farming and fishing, are moderately affected by the environmental change in Sundarban. Both migrants and non-migrants perceive more or less equal environmental risk in their respective activities. The impact of environmental change, although existing, is subsumed under socio-economic conditions. It could be considered as an associate driver working in tandem with other predominant factors. Meanwhile, it also evokes to address the role of other factors like assets, capabilities, structure and process (i.e. governmental and non-governmental policy and programmes), nexus with environmental change, individual effects of each factor on migration controlling the other etc. These are needed to explore with empirical rigour the environmental migration vividly.

Notes

1 Trash fish or rough fish are those that are less desirable to the catchers in a region. Environmentalists suggest that the production of rough fish is increased due to the climate change.
2 Catch per Unit Effort (CPUE) refers to the total amount of fish caught by a boat. Its unit is kg/haul.

References

Black, R., Arnell, N. W., Adger, W. N., Thomas, D. and Geddes, T., 2012. Migration, Immobility and Displacement Outcomes following Extreme Events. *Environmental Science and Policy*, Vol. 27, pp. S32–S43.

Bureau of Applied Economics & Statistics (BAE&S), 2004 to 2011. *District Statistical Handbook, South 24 Parganas*. The Government of West Bengal. Kolkata: New Secretariat Buildings. CD-ROM.

Bureau of Applied Economics & Statistics (BAE&S), 2010. *District Statistical Handbook, South 24 Parganas*. The Government of West Bengal. Kolkata: New Secretariat Buildings. CD-ROM.

Camber, R., and Conway, G. R., 1991. Sustainable Rural Livelihoods: Practical Concepts for 21st Century. IDS Discussion Paper 296. From http://opendocs.ids.ac.uk/opendocs/bitstream/handle/123456789/775/Dp296.pdf?sequence=1. Accessed on 25 July 2013.

Census of India, 2011. Census Tables, Office of the Registrar General and Census Commissioner, Ministry of Home Affairs, Government of India. CD-ROM.

DFID (The Department for International Development), 2001. *Sustainable Livelihoods Guidance Sheets*. London: DFID.

Falco, S. D., et al., 2011. Estimating the Impact of Climate Change on Agriculture in Low-Income Countries: Household Level Evidence from the Nile Basin, Ethiopia. *Environmental Resource Economics*, Vol. 52, No. 4, pp. 457–478.

Farauta, B. K., et al., 2011. *Farmers' Perceptions of Climate Change and Adaptation Strategies in Northern Nigeria: An Empirical Assessment*. Nairobi, Kenya: African Technology Policy Studies Network.

Ghosh, A., 2012. Living with Changing Climate: Impact, Vulnerability and Adoption Challenges in Indian Sundarbans. Centre for Science and Environment, New Delhi. From http://cseindia.org/userfiles/Living%20with%20changing%20climate%20report%20low%20res.pdf. Accessed on 20 December 2013.

Hazra, S., et al., 2010. Temporal Change Detection (2001–2008) Study of Sundarban. From www.iczmpwb.org/main/pdf/ebooks/WWF_FinalReportPDF.pdf. Accessed on 23 January 2013.

IPCC, 2007. *Impacts, Adaptation and Vulnerability, Contribution of Working Group II to the Fourth Assessment Report of the Intergovernmental Panel on Climate Change*. Cambridge: Cambridge University Press.

Juana, J. S., Kahaka, Z. and Okurut, F. N., 2013. Farmers' Perceptions and Adaptations to Climate Change in Sub-Sahara Africa: A Synthesis of Empirical Studies and Implications for Public Policy in African Agriculture. *Journal of Agricultural Science*, Vol. 5, No. 4, pp. 121–135.

Leiserowitz, A., 2006. Climate Change Risk Perception and Policy Preferences: The Role of Affect, Imagery, and Values. *Climatic Change*, Vol. 77, pp. 45–72.

Mistri, A., 2013. Migration and Sustainable Livelihoods: A Study from Sundarban Biosphere Reserve. *Asia Pacific Journal of Social Sciences*, Vol. 5, No. 2, pp. 76–102.

Mistri, A. and Das, B., 2015. Environmental Legislations and Livelihood Conflicts of Fishermen in Sundarban, India. *Asian Profile*, Vol. 44, No. 3, pp. 389–400.

Mitra, A., et al., 2009. Observed Changes in Water Mass Properties in the Indian Sundarbans (Northwestern Bay of Bengal) During 1980–2007. *Current Science*, Vol. 97, No. 10, pp. 1445–1452.

Myers, N., 2005. "Environmental refugees, an emergent security issue", Paper presented to the 13th Economic Forum, Prague, Czech Republic, 23–27 May 2005.

Singh, O. P., 2007. Long-Term Trends in the Frequency of Severe Cyclones of Bay of Bengal: Observations and Simulations. *Mausam*, Vol. 58, No.1, pp. 59–66.

Singh, O. P., Khan, T. M. A. and Rahman, Md. S., 2001. Has the Frequency of Intense Tropical Cyclones Increased in the North Indian Ocean? *Current Science*, Vol. 80, No. 4, pp. 575–580.

Slovic, P., 1987. Perception of Risk. *Science*, Vol. 236, No. 4799, pp. 280–285.

Slovic, P. and Peters, E., 2006. Risk Perception and Affect. *Current Directions in Psychological Science*, Vol. 15, No. 6, pp. 322–325.

Stern, N., 2006. *The Economics of Climate Change: Stern Review*. Cambridge, UK: Cambridge University Press.

UNDP, 2007. *Fighting Climatic Change: Human Solidarity in a Divided World, Human Development Report 2007/2008*. Geneva: United Nations Development Programme.

9 Gender processes in rural outmigration and socio-economic development in the Himalaya[1]

Prakash C. Tiwari and Bhagwati Joshi

The Himalaya represents one of the tectonically most unstable, ecologically fragile, economically underdeveloped and most densely populated mountain ecosystems on the planet, and is inhabited by some the poorest and most marginalised people of the world (ICIMOD, 2011). It presents geographically, geologically and culturally unique landscapes which constitute the source of a range of ecosystem services that sustain hundreds of millions of people both upstream and downstream (Woodroffe et al., 2006; Macintosh, 2005). Constraints of terrain and climate impose severe limitations on the carrying capacity of the natural resource base and on the effectiveness of infrastructural services and facilities. As a result, subsistence agriculture constitutes the main source of rural livelihood and food even though the availability of arable land is severely limited, and agricultural productivity is considerably low (Tiwari and Joshi, 2013). Due to the inherent limitation and risks descending from subsistence economy, a large proportion of adult male population outmigrates in search of livelihood (Maithani, 1996). In the recent past, a variety of changes has emerged in traditional resource utilisation patterns, mainly in response to population growth and resultant increased demand of natural resources. As a result, critical natural resources, such as forests, pastures and rangelands, have deteriorated and depleted steadily and significantly leading to their conversion into degraded and non-productive lands in the region over the last 20 to 30 years (Tiwari and Joshi, 2012a).

As a result, the productivity of rural ecosystem has declined considerably and the livelihood security of rural poor has been adversely affected, leading to environmental instability and community unsustainability in the region. The rapid urbanisation has exerted sharply accentuated pressures on local subsistence economy through depletion of land, water, biodiversity and forest resources collapsing conventional production system and increasing community vulnerability to livelihood and food insecurity and increased risks of natural disasters (ICIMOD, 2009, 2010; Tiwari and Joshi, 2012b; World Bank, 2009; Lonergan, 1998). Economic globalisation has further increased the vulnerability of mountain communities to environmental risks through exploitation of natural resources even in remote and inaccessible areas and seems to have further strengthened poverty imbalances between highlands and lowlands (UNDP, 2010; Hassan et al., 2005; Huddleston and Ataman, 2003; Stromquist, 2002; Verma, 2001). Moreover, changing climatic conditions have stressed the Himalayan ecosystem through higher mean annual temperatures and melting of glaciers and snow, altered precipitation patterns and more frequent and extreme weather events which are likely to intensify the impacts of other natural as well as socio-economic drivers of change (IPCC, 2014, 2007; UNEP-WCMC, 2002). This has augmented male outmigration, and consequently increased the hardships of rural women and deteriorated their quality of life (Tiwari and Joshi, 2012b; Brody et al., 2008; CIDA, 2002).

The increasing trend of draining away human resource from the mountains has serious implications not only for economic development, but also for the enrichment of socio-cultural life in the region. This led to the feminisation of natural resource management process and of agriculture in the Himalaya (ICIMOD, 2010). Consequently, Himalayan women have often been designated as 'primary resource developers', as the burden of living under difficult mountain conditions falls mainly on them: women have to bear the drudgery of scrounging for all primary natural resources, including fuelwood and fodder collection from shrinking forests, water fetching from increasingly long distances, besides taking care of agriculture, livestock, children and aged members of the family (Agarwal, 2010; Sharma and Banskota, 2006). Still, they have very limited ownership of and access to natural resources and limited opportunities of involvement in decision-making processes (Joshi and Tiwari, 2014; Byers and Sainju, 1994; Kelkar, 2005). On the other hand, there is severe shortage of adult male labour to work in agriculture and in other sectors of the rural economy, which further acts as drag on agricultural productivity and retards the process of socio-economic development in the mountains. The cumulative impact of all these physical, socio-economic and cultural constraints is grinding poverty, hardship, constant fear of insecurity, a feeling of helplessness and complete dependence on outside help. The environment thus created is not conducive for attaining the goals of sustainable mountain development (Maithani, 1996). However, women experience these changes differently and disproportionately and respond to them in varying manner because of socially constructed gender relations and of the environmental sensitivity of mountain ecosystems. Women make use of their critical traditional knowledge and experience in natural resource management and adapt agricultural and food systems to multiple drivers of environmental change, including climate change, globalisation and economic processes, outmigration and land-use changes in mountain environments, which helped women to become important agents of sustainable mountain development (ICIMOD, 2011). The main objective of the study is to analyse the trends and drivers of male youth outmigration and interpret their linkages to the socio-economic status and development of rural women with an empirical study of the Upper Kosi Catchment located in the newly carved Himalayan State of Uttarakhand in India.

The conceptual background

Gender refers to the range of socially constructed roles, behaviours, attributes, aptitudes and relative power that are established by social traditions, customs and conventions and associated with being female or male in a given society (Esplen, 2009). These social traditions, customs and conventions often restrict women's access to crucial natural assets, such as land, water and forests as well as strategic resources, such as capital, information, legal rights and ownership and education (Khosla and Pearl, 2003). As a result, women and men experience environmental changes in different ways and magnitude (Aguilar, 2006). In the Himalaya women experience environmental changes acutely and unevenly, facing discrimination at the intersection of gender, class, caste, ethnicity and other social domains of difference and inequity. As a result women are particularly dependent on climate-sensitive livelihood options, such as subsistence agriculture (Fisher, 2006; Khosla and Pearl, 2003). The gender inequalities in the distribution of and access to natural resources, assets and opportunities severely restrict women's options and choices in the face of environmental and climatic changes (Masika, 2002; Dennison, 2003). For example, restrictions on the ownership of land and other natural resources reduce women's access to productive land and availability of food, whereas the lack of financial resources restrict diversification of their livelihoods and access to food (Clancy and Skutsch, 2003; Department for International Development, 2007).

Migration is a social process embedded within a variety of other social practices. Moreover, gender-influenced cultural expectations, policies and institutions play a critical role in determining the causes and consequences of migration (Jolly and Reeves, 2005). The steady depletion of natural resources has accelerated the trends of male outmigration which have enhanced women's roles and responsibilities, and increased their workload both in the agricultural and domestic sectors (Mitchell et al., 2007; Agarwal, 2009; Baniya et al., 2005). Moreover, climate change and uncertainties associated with the frequency and intensity of extreme weather events have reduced agriculture potential and disrupted 'ecosystem services' of forests, biodiversity, freshwater and soils, adversely affecting rural livelihoods rendering rural ecosystem more vulnerable to environmental changes, particularly to climate change (Sever, 2005). These changes have further increased the trends of both short-term and long-term outmigration of males from the region (Reuveny, 2007; Ramachandran, 2006).

The framework of sustainable livelihood approach has been modified and successfully used in a wide variety of rural development programmes (e.g. World Bank, 2005; Yaro, 2006). The sustainable livelihood framework is based on the concept that the relative availability of various 'capital assets' shapes livelihood options. The assets include: (i) natural (i.e. land, water, forests); (ii) social (i.e. community, family, social networks); (iii) political (i.e. participation, empowerment – sometimes included in the 'social' category); (iv) human (i.e. education, labour, health, nutrition); (v) physical (i.e. roads, clinics, markets, schools, bridges); and (vi) economic (i.e. jobs, savings, credit). The relative availability of these livelihood assets is determined by individual and household actions as well as broader socio-economic structures and processes and political frameworks. However, livelihood frameworks and strategies are further determined by households' vulnerability to shocks and stresses, such as climate change and extreme weather events. Socio-culturally defined gender roles shape both perceived values of 'assets' (e.g. differential value placed on men's and women's human capital) as well as actual household and individual use of the assets themselves (e.g. natural resource collection perceived as women's work) (World Bank, 2005).

In the Himalaya, the constraints of terrain and climate impose severe limitations on the productivity of natural resources, and consequently subsistence agriculture constitutes the main source of rural livelihood even though the availability of arable land is severely affected and agricultural productivity is low. The combined hardship and adversity of mountain living conditions have trained local women to develop traditional agriculture as a means of sustainable livelihood and to respond radically to the forces that were exploiting their natural resources and undermining their livelihoods (Mehta, 2007). Active participation of rural women in the conservation and management of their natural resources provided able leadership to a series of environmental conservation movements against exploitative state policies in Uttarakhand Himalaya. The famous 'Chipko Movement' (hugging the trees movement) of Uttarakhand – a distinctly non-violent grassroots movement – was organised and lead by illiterate rural women against the suppressive state forestry policy in the 1970s.

The present work is an attempt to demonstrate the hypothesis that male outmigration in the Himalaya has not only provided stability to rural economies in terms of remittances, but also marginally improved women's access to education, development opportunities, leadership, decision-making power, natural resource management and growing market from local to global levels (Ratha and Xu, 2008). These changes are contributing towards social, economic and political empowerment of rural women and providing them opportunities to engage in decision-making process from family to village levels. Furthermore, women have developed critical traditional knowledge to understand, visualise and respond to environmental changes including climate change. Women's empowerment refers to women's sense of self-worth, right to have and determine choices, right to have access to opportunities and resources; their right to have the power to control their own lives, both within and outside the home; and their ability to influence the direction of social change to create a more just social and economic order, nationally and internationally (United Nations, 2006; Oxaal and Baden, 1997; Malhotra et al., 2002). Nevertheless, very little work has so far been carried out on gender and migration in the context of environmental change, gender and mainstreaming and women's empowerment in the region. The main objective of the present study is to investigate the impacts of male outmigration through the interpretation of primary data and empirical observations in the region.

Study area and research methods

The Upper Kosi Catchment encompasses a land surface area of 10,794 km^2 (10,794 ha) in the densely populated Lesser Himalayan ranges of the state of Uttarakhand in India. The total population of the region is 16,080 organised in 62 villages. The watershed is one of the most densely populated and agriculturally colonised valleys of Kumaon Himalaya with an average population of 149 persons/km^2. The major land-use categories are forests and cultivated land as they respectively share 71 per cent and 26 per cent of the total geographical area of the catchment. Subsistence agriculture with animal husbandry as its natural ally constitute the main sources of livelihoods for more than 75 per cent of the population, although the availability of arable land is severely limited and agricultural productivity is significantly low in most parts of the catchment, particularly in the villages located in the upper slopes. More than 87 per cent of operational land holdings are of less than 1 ha and the availability of cultivated land is only 0.17 ha per capita, resulting in high dependency on limited arable land and low agricultural productivity. This compels a large proportion of adult male population to outmigrate the region in search of livelihood and employment. The entire Upper Kosi Catchment has been divided into four micro-watersheds for comprehensive study of various research parameters included in the study. The micro-watersheds include: (i) North Kosi Micro-watershed, (ii) South Kosi Micro-watershed, (iii) East Kosi Micro-watershed and (iv) West Kosi Micro-watershed.

Methodology included generation and analysis of primary socio-economic data pertaining to (i) trends and drivers of rural male youth outmigration; (ii) interpretation of inter-linkages among the parameters of subsistence agricultural economy, resource depletion, local production systems, traditional livelihood, outmigration and women and (iii) analysis of the impact of depleting natural resource base and climate change on conventional rural livelihood, and appraisal of its impact on rural outmigration and the socio-economic status of rural women. The information and data required for the study have been generated and collected from various primary and secondary sources. The primary information has been generated through intensive field surveys and mapping, observations, monitoring and socio-economic surveys. The information pertaining to outmigration was collected from Village Register of each of the 62 villages of the catchment, and also through household surveys. In order to develop the estimates of production and demand of natural resources comprehensive socio-economic surveys were conducted in the year 2010 using exclusively designed village and household schedules. General information about forest, agricultural land and water resources have been collected from each of the 62 villages using village schedule by interviewing the head and members of Gram Pradhan (Village Council). The information was generated through household surveys conducted in all the villages, with respect to production and demand of food, fuelwood and fodder; issues associated with increasing trends of outmigration; and factors leading to transformation of community resource utilisation structure and depletion of natural resources and their impact on traditional rural livelihood, resource collection distances, work load on women, community health, water availability and utilisation, food consumption pattern and impact of natural disasters. The sample size covered 33 per cent of the total households (out of total 2,197 households) selecting respectively women-headed households (25 per cent), households below poverty line (as classified by the Government of Uttarakhand State) (40 per cent), households solely dependent on agriculture (15 per cent) and households dependent on agriculture and other means of income (20 per cent).

Drivers and trends of migration

The local communities are traditionally forest-dependent. Forests are therefore an integral component of their livelihood, economy, culture and history. Hence, the traditional resource-use structure is closely interlinked with forests, pastures, livestock and cultivated land, and forest-based subsistence agriculture constitutes the main source of rural livelihood for more than 75 per cent of the population. In order to preserve soil fertility level and productivity of land under sustained cropping, in such an agro-ecosystem, there must be a net transfer of energy from forests to arable land. This flow of energy from forest to cultivated land is mediated through livestock, usually in the form of fodder of stall-fed cattle whose manure and labour form the main source of energy to the Himalayan agricultural system. In order to meet the energy requirement of 1 ha of arable land nearly 5–12 ha of well-stocked forest is required, and on an average one unit of agricultural production involves nine units of energy from surrounding forests (Singh et al., 1984). However, due to land-use intensification the availability of forests to per ha cultivated land ranges between only 0.10 and 0.22 in different micro-watersheds.

A minimum of 0.2 ha per capita arable land is necessary for practising agriculture on sustainable basis in the Himalaya (Ashish, 1983), whereas, the average availability of cultivated land is merely 0.16 ha/person in the catchment. Land holding size is very small and not adequate to carry out agriculture on a sustainable basis both in terms of productivity and food security in the region (Ashish, 1983). The interpretation of primary data revealed that approximately 87 per cent of the households belong to the category of small farmer with land holding size less than 1 ha, 7 per cent fall in the category of medium-small farmer with land holding between 1–2 ha, 5.5 per cent are medium farmers with land holding between 2–3 ha and only 0.5 per cent families have been categorised as large farmers with land holding size of more than 3 ha. The small size of land holdings is increasing biotic stress on the natural ecosystem and leading to land-use intensi-fication and depletion of the natural resource base (Tiwari and Joshi, 2012a, 2012b). The rapid land-use changes and decline in forest area have disrupted ecosystem services resulting in decline in agricultural productivity and considerable loss of livelihood in traditional agricul-ture and forestry sectors. Moreover, the changing climatic conditions have stressed Himalayan agricultural and livelihood systems through higher mean annual temperatures and melting of glaciers and snow, altered precipitation patterns and hydrological disruptions and more frequent and severe extreme weather events. The analysis of rainfall data collected from local weather station brought out the facts that the amount of rainfall and the number of rainy days have declined respectively by 52 per cent and 34 per cent, and the incidences of high intensity rainfall and droughts have increased during last ten years. The present study clearly indicated that these changes have disrupted hydrological systems and reduced the availability of water resulting in frequent crop failures, decline in irrigation potential (25%), decreased agricultural productivity (26%), loss of rural livelihood (34%) in tra-ditional rural sectors and increase in food deficit level (Rawat, 2009; Malhotra and Schuler, 2005). The interpretation of primary data generated during the field surveys revealed that the different micro-watersheds of the catchment are currently facing food deficits between 35 per cent and as much as 75 per cent with an overall food deficit of 65 per cent.

However, mountain communities have developed mechanisms to adapt to changing conditions (ICIMOD, 2010; Leduc and Shrestha, 2008; UNEP, 2004). Migration of male youth is one of the important adaptive measures to constraints of subsistence economy and changing environmental conditions and associated natural and socio-economic risks all across the mountain regions of the world (Sherpa, 2007; ICI-MOD, 1999). It was emerged during the intensive discussion held with the local people that depletion of natural resource base, climate change and consequent loss of livelihood opportunities and decline in food production have accelerated the process of male outmigration in the region. Table 9.2 shows that 21,496 persons have migrated from the region from 2001 to 2010, out of which 81.48 per cent were educated minimum up to the level of high school. Out of the total migrants, 97 per cent were males and 27.79 per cent migrated permanently while 72.21 per cent migrated on a temporary basis (Table 9.2). The study observed that the male outmigration has consistently increased in recent years. In 2001 only 701 people migrated from the region, whereas the number of outmigrants increased to 5,511 in ten years thus registering an overall

Table 9.1 Trends of rural outmigration during 2001–10

Years	Total migrants	% change
2001	701	–
2002	795	13.41
2003	1,007	26.67
2004	1,295	28.60
2005	1,491	15.13
2006	1,521	02.01
2007	1,609	05.78
2008	2,975	84.90
2009	4,591	54.32
2010	5,511	20.04

Source: Field surveys 2010 and Village Record

Table 9.2 Pattern of rural outmigration during 2001–10

Micro-watersheds	Total migrants	Permanent migrants	Temporary migrants	Educated migrants	Uneducated migrants
North Kosi	4,076	37.00	63.00	75.21	24.79
East Kosi	4,560	21.55	78.45	81.55	18.45
West Kosi	5,230	37.11	62.89	77.59	22.41
South Kosi	7,630	15.51	84.49	91.57	08.43
Total	21,496	27.79	72.21	81.48	18.52

Source: Field surveys 2010

increase of nearly 686 per cent during the period (Table 9.1). The interpretation of primary data revealed that poverty, decline in agricultural productivity, loss of livelihood opportunities and increasing incidences of natural disasters have been the most pressing factors for increasing trends of male outmigration in the watershed (Table 9.3).

The development potential in traditional subsistence agriculture-livestock system has declined mainly due to population growth, low carrying capacity and depletion of natural resource base and consequent loss of ecosystem services in the region. Moreover, rapidly changing climatic conditions have stressed the Himalayan agricultural system through higher mean annual temperatures, altered precipitation patterns, more frequent and extreme weather events, which are increasing the vulnerability of rural communities to food and livelihood insecurity (Rasul, 2014; Singh et al., 2010; Sharma et al., 2007; Sah and Batarya, 2006;

Table 9.3 Drivers of male outmigration: a community perception

Drivers of rural outmigration	Community response (% of 4,000 persons)
Poverty	37
Lack of Livelihood Opportunities	27
Decline in Agricultural Productivity	21
Extreme Weather Events and Natural Disasters	11
Landless and Marginalized	02
Lack of Services and Facilities	01

Source: Field surveys 2010 and Village Record

Singh and Bengtsson, 2004; Beniston, 2003; Tiwari and Joshi, 2012a, 2012b; ICIMOD, 2011). Consequently, agricultural productivity is declining and livelihood opportunities in traditional sectors are decreasing. Since there is almost lack of other viable means of rural livelihood and employment in the region partially owing to ecological constraints and partially due to inappropriate and ineffective process of development, a considerably large proportion of rural population depends on subsistence agriculture. The people are carrying out agriculture in most compelling circumstances as the availability of arable land is severely limited (less than 0.2 ha/person), agricultural productivity is low and there are no other viable means of livelihood in the region. In order to raise food production from severely limited agricultural land farmers try to increase the number of crops grown in an agricultural year which enhances cropping intensity but ultimately stresses limited agricultural resources and declines agricultural productivity (Maithani, 1996).

Impacts on social and economic development of women

It was observed that besides providing the economic stability and security in terms of remittances to the Himalayan communities male outmigration is now also contributing towards the socio-economic development of rural women (Hoermann and Kollmair, 2009; Kollmair et al., 2006). The findings of the present works which are presented in the proceeding sections substantiate these observations.

Educational empowerment of women

Male outmigration has marginally contributed towards increasing the level of female education in many villages of the region (Banerjee et al.,

Table 9.4 School dropout rates of girl children and level of female education during 2001–10

Micro-watersheds	Girl child school dropout rates (%)			Level of female primary education (%)		
	2001	*2010*	*% change*	*2001*	*2010*	*% change*
North Kosi	37	31	06	39	55	16
East Kosi	41	29	12	55	63	08
West Kosi	55	35	20	55	69	14
South Kosi	51	25	26	57	71	14
Total	46	30	16	52	65	13

Source: Field surveys 2010

2011; Dang et al., 2009; Nautiyal, 2003). The interpretation of primary data generated during the field surveys revealed that the school dropout rate of girl children has decreased by 6 per cent to 26 per cent with an average decline of 16 per cent in the entire watershed during the last ten years (Table 9.4). Similarly the level of primary education among women also registered an increase ranging between 8 per cent and 16 per cent in different micro-watersheds with an overall increase of about 13 per cent in the catchment during 2001–2010 (Table 9.4). This information was generated for both migrant and non-migrant households, and authors found that the school dropout of girls was quite lower and the proportion of girls enrolled at primary schools in the watershed was much higher in migrant households. This clearly indicates that male outmigration besides contributing towards economic security to their families also empowers rural women to take decisions for their educational and social empowerment. The women in many villages of the catchment have been successful in creating social space and opportunities in the process of critical social and economic decision making both at household as well as at village levels. The authors observed during village and households surveys that the important matters in which women have been able to influence the decision-making process include selection of agricultural crops and sale of agricultural and livestock products at family level, and in determining village-level developmental priorities. The similar qualitative empirical observations have also been reported in number of other studies (Sidh and Basu, 2011; Mamgain, 2003). It was observed that the level of women's educational and social empowerment is much higher in female-headed families than in traditional male-headed households in the region.

Table 9.5 Level of social empowerment of rural women

Micro-watersheds	Participation in grassroots development institutions (% women)			Participation in natural resource management institutions (% women)		
	2001	*2010*	*% change*	*2001*	*2010*	*% change*
North Kosi	05	17	12	10	21	11
East Kosi	07	15	08	05	16	11
West Kosi	10	25	15	08	17	09
South Kosi	15	25	10	10	15	05
Total	09	21	12	08	17	09

Source: Field surveys 2010

The educational awareness and empowerment are motivating women to participate and take leading roles in grassroots institutions – Gram Panchayats (Village Councils) – democratic and constitutional institutions created in every village of India for the decentralisation of developmental governance. Table 9.5 makes it clear that the women's participation in local governance has increased between 8 per cent and 15 per cent in different micro-watersheds with an average increase of 12 per cent during last ten years. Since, women's participation is increasingly on the development agenda of government and non-governmental organisations, during the recent decades there has been growing impetus among governments and civil societies for political empowerment of women particularly in local developmental governance structure. The Constitution of India mandates the reservation of a minimum of one-third of seats for women both as members and as Sarpanch (Chairperson of Village Council) in Panchayati Raj Institutions (PRIs). The constitution also delegated Panchayats the responsibility of furthering the agenda of socio-economic development and social equality. More recently in 2009, the Government of India increased participation for women in PRIs from 33 per cent to 50 per cent, and the Government of Uttarakhand also implemented this through similar legislation. The empirical observations indicated that these constitutional provisions increased the involvement of rural women in grassroots governance system. However, these constitutional provisions did not facilitate the effective participation of rural women in local governance, as women representatives continue to face several institutional and social constraints pertaining to the capacity of local governance structures to implement reforms, institutionalisation of accountability systems, decentralisation of functions and level of education and respect for women in PRIs. However, in several villages in the study area it was observed that educated women not only provide effective leadership as Sarpanchs to these local governance institutions but also influence the decision-making process as members of Village Councils. The above-mentioned findings clearly indicate that on the one hand male outmigration has developed awareness among the women about the importance of social, economic and political empowerment, and on the other it has made them to realise the significance of education in their lives.

Political empowerment of women

In Uttarakhand Himalaya existed probably the first participatory form of statutory resource management institution in the world with a long history of about 90 years in the form of forest panchayat or van panchayat (village forest council).[2] In fact, forest panchayats in Uttarakhand emerged out of conflicts and compromises that followed the settlements and reservations of forests in the region during colonial period in the early 20th century, and the first state-approved forest panchayat came into existence back in 1931 in district Almora (Mukherjee, 2003). Thus forest panchayats are a localised form of community forestry institutions in Uttarakhand Himalaya that are locally elected democratic bodies or voluntary groups of local people that govern the local forests with a view to fulfil the needs of local people for forest produce in a sustainable and equitable manner (Mishra et al., 2008).

However, this system of forest governance collapsed after independence mainly due to more administrative control of civil administration and forest department over the Forest Panchayats. However during recent years, the forest department tried to revive Forest Panchayats across the Uttarakhand State by making them more democratic, self-governing, autonomous in decision making and community-oriented particularly through active involvement of poor and marginalised people, specifically of women. The participation of women in Forest Panchayats increased between 5 per cent and 11 per cent in different micro-watersheds of the catchment during the last one decade (Table 9.5). The active involvement of women in Forest Panchayats linked forest conservation with livelihood-based natural resource management, income generation from forest biomass residue and an increased flow of financial resources for promoting a green economy by strengthening institutional governance of forests. It was observed that Forest Panchayats are also contributing towards conserving ecosystem services including carbon sequestration, which has acquired a significant value in the context of climate change.

These local forestry institutions contributed significantly not only towards the conservation of forests, but also towards linking forest governance with betterment of quality of life in rural areas, empowerment of women in natural resources, capacity building of grassroots institutions and above all integrating forest management with climate change mitigation and adaptation, and global carbon markets. It was observed that in several cases a participatory forest management system has enhanced carbon sequestration capacity of forest ecosystem through increase in forest cover and strengthening community resilience by improving traditional adaption practices in water, biodiversity and livelihood sectors (Singh, 2007; Kant, 2011). The Forest Panchayats have been successful in interlinking changing local needs, national development priorities and international environmental concerns at the local level through contributing towards conservation of forests and biodiversity, watershed restoration, rural livelihood improvement and enhanced carbon sequestration (ICIMOD, 2011; Shyamsundar et al., 2011). The program also presented a good example of the local implementation of the global agenda on environmental governance and sustainable development (Shyamsundar et al., 2011). It was observed during field surveys in the present study that in most of the cases the Forest Panchayats headed by women of migrant families are better managed in terms of increasing forest area, utilisation of forest products, conservation of water resources, access to resources and benefits and community involvement in decision making than those headed by men (Singh, 2007). Besides, women are involved actively in other community-based natural resource management institutions, particularly 'Swajal' – a village-level participatory rural water governance system.

The economic empowerment of women

The economic security through remittances and improvement in educational level also helped rural women in developing entrepreneurship and seizing the emerging developmental challenges. Economic globalisation, decentralised governance mechanisms, rapid urban growth and development of information technology and communication system have sensitised mountain communities for capturing the potential of these drivers of transformation. During the recent years, with the improved road connectivity and growth of tourism, a significant proportion of educated women have entered into small-business sectors

Table 9.6 Level of Economic Empowerment of Rural Women

Micro-watersheds	Women Running Rural Economic Enterprises (% Women)			Women Operating Bank Account (% Women)		
	2001	2010	% change	2001	2010	% change
North Kosi	02	07	05	02	15	13
East Kosi	01	11	10	03	11	08
West Kosi	02	05	03	05	09	04
South Kosi	07	10	03	10	15	05
Total	03	08	05	05	13	08

Source: Field surveys 2010

and established handicraft outlets, guest houses, restaurants, Internet cafes, travel agencies, fruit and vegetable collection centres, beauty parlours and goods forwarding centres along the road. Besides, a considerable proportion of rural women are operating their bank account in nearby branches of rural and other banks (Table 9.6). Table 9.6 shows that the women-owned business enterprises have increased 3–10 per cent in different micro-watersheds of Upper Kosi Catchment during a small period of one decade. This is bringing economic autonomy and empowerment to rural women which are clearly indicated by the proportion of women holding a bank account that has increased from 5 per cent to 13 per cent during 2001–2010.

Male outmigration and women's adaptation to climate change

The increasing trends of male outmigration have provided women with the opportunity of perceiving, experimenting with and adapting to changing environmental conditions, particularly climate change. Being the primary resource developers women have developed critical traditional knowledge to understand, visualise and respond to environmental changes including climate change. The study revealed that women have not only been very well able to observe the impacts of rainfall variability on local agroecosystem, but also already started responding to the recognised changes in rainfall and resultant depletion of water resources using their traditional resource management experience and indigenous knowledge (Joshi and Tiwari, 2014). The interpretation of data generated through primary surveys conducted in ten villages (out of total 62 villages of Upper Kosi Catchment) brought out the interesting facts about the traditional coping and response mechanism that mountain indigenous women have evolved to adapt their agricultural resource development processes to climate change in the region. The study made it clear that women, particularly women-headed households of the region, have been using several adaptation measures to maintain a minimum level of agricultural productivity under changing rainfall conditions and declining rainfall and rainy days (Tiwari and Joshi, 2014). The important adaptation measures being used by rural women are presented in Table 9.7 and explained below:

Table 9.7 Women's traditional knowledge and adaptation to climate change

Villages	Total house-holds	Women-headed house-holds	% women-headed households practising adaptive measures				
			Rain-water harvesting	Traditional water conservation	Change in cropping pattern	Change in crop rotation	Wasteland development
Kantali	65	21	30	21	31	19	09
Bheta	71	32	19	45	17	28	07
Bilori	57	27	17	37	23	27	17
Chanoli	55	19	29	25	09	25	15
Dhalaur	97	47	27	15	11	17	07
Hatyura	44	17	21	21	35	21	08
Kwerali	56	25	35	27	31	15	14
Lod	51	15	27	35	19	23	11
Lewsal	75	31	19	15	05	23	10
Mala	72	27	25	27	05	14	11
Total	643	261	25	27	19	21	11

Source: Field surveys 2010

- Nearly 27 per cent of women have replenished their water sources through water conserving forestry and horticultural practices.
- About 19 per cent of women-headed households cultivated crops which are drought resistant and need less water.
- Approximately 25 per cent of women cultivators sustainably managed depleting water resources by evolving and practising locally effective rainwater collecting measures based on their traditional water resource management system.

- Nearly 21 per cent of women-headed households altered traditional cropping patterns and adjusted crop rotation.
- In order to increase food production under rainfall uncertainty conditions women-headed households (11%) cultivated abandoned agricultural land, and 27 per cent tried to successfully relocate their farming system through improved crop rotation.
- It was observed through empirical studies that women-headed households have abandoned their less productive and marginal and sub-marginal agricultural land and started small business along the roads (7% women-headed households).
- About 5 per cent of agriculturally marginalised women families outmigrated their villages and settled with their relatives in other parts of the country.
- Some poor women-headed households accommodated to climate change by lessening the consumption of less productive and expensive food crops such as lentils and rice.

The limiting factors

However, the increasing trends of male outmigration have not only affected the life quality of rural women through increasing their responsibilities and workload of resource development, but also eroded the rich traditional knowledge which rural communities, particularly women have developed through their long experimentation with nature and changing natural conditions. Women in Himalaya are particularly vulnerable to the impacts of environmental changes due to skewed power relations and unfair cultural and social norms, as a result, they are often disadvantaged in terms of power relations and access and ownership to natural resources (Mamgain, 2003). Furthermore, in Himalaya, women traditionally have sole responsibility of household tasks as well as outdoor resource development and collection liability which drives them to interact more with the surroundings that make them more vulnerable to risks of environmental changes, particularly climate change (Malla, 2008; Singh and Svensson, 2010). Women are more vulnerable during disasters because they have less access to resources, information, education and ownership of physical assets, and are victims of the gendered division of labour (Ramachandran, 2006). Moreover, women suffer more during the post-disaster phase due to the sharp increase in workloads because of their multiple roles and collapse of their traditional income-generating activities (Uniyal, 2013; Sherpa, 2007; Ramachandran, 2006; Joshi and Tiwari, 2014; ICIMOD, 2010). Moreover, increasing trends of rural male outmigration have increased women's roles and responsibilities, enhanced their workload both in agricultural and in domestic sectors. Despite the positive changes that are transforming the lives of rural women, women's access to education, information, communication and health facilities and participation in decision-making process still remain at a very low level in most of the villages (Joshi and Tiwari, 2014). This strengthens feminisation of rural poverty which further carries the risks of marginalisation, exclusion from decision making and reduced access to natural and financial resources (Sherpa, 2007). The increasing trends of male outmigration inadvertently created an emptiness in the mountain societies putting extra responsibilities on women (Sherpa, 2007). The draining away of productive human resources from mountains has serious implications not only for the economic development, but also for the enrichment of socio-cultural life in the region (Maithani, 1996). Women, being natural resource developers, possess undocumented indigenous knowledge, and their contribution towards preserving the mountain cultures and natural resources is highly significant (Uniyal, 2013; Gupta, 2007; Mukherjee, 2003; Mamgain, 2003; ICIMOD, 1999; UNDP, 2010).

The increasing trends of male outmigration and depletion of the natural resource base have cumulative impacts on the quality of life of rural women across the Himalaya. The studies carried out by the authors indicate that due to depletion of forest and water resources the average travel distance involved in the collection of fodder, fuel-wood and water has increased on an average between 0.5 km and 1.5 km in different micro-watersheds of the catchment during the last ten years (Joshi and Tiwari, 2014). As a result, the workload of agricultural and livestock sectors on women has increased (30%), and the time available with women for personal and child care has reduced (25%) during 2001–10 (Tiwari and Joshi, 2014b). The depletion of water resources has not only stressed the water supply system but also reduced the productivity of traditional agricultural system undermining community health through decreasing availability of water for domestic purposes and reducing the quantity as well as quality of food affecting mainly women of all age groups (Tiwari and Joshi, 2012b, 2014a, 2014b). It was learned from recent devastating natural disasters in Uttarakhand Himalaya, possibly induced by climate change, that physiological vulnerabilities, socio-cultural and economic marginalisation and gender stereotypes still make the differential level of impacts on mountain communities (Uniyal, 2013; Kumar, 2013). As a result, the nature and magnitude of the impacts of natural disasters on men and women are completely uneven during as well as after the disaster

(Mehta, 2007). The women constitute the highest proportion of population affected by natural disasters primarily due to lack of preparedness, information and exposure. The outmigration of males during the post-disaster phase, often forces women to shoulder the productive, reproductive and community responsibilities, while still suffering from the agony caused by the disaster (Nellemann et al., 2011).

Conclusion

In the Himalaya, the constraints of subsistence agricultural economy and absence of other viable means of rural livelihood compel a large proportion of adult male population to outmigrate the region in search of livelihood and employment leading to feminisation of local resource development process, agriculture and poverty. This not only increases the roles, responsibilities and workload of rural women but also deteriorates the quality of life of rural women. Moreover, the ongoing process of exploitation and depletion of critical natural resources, such as land, water, forests and biodiversity and climate change, increased not only the vulnerability of rural communities to water, food, livelihood and health insecurities but also the susceptibility of mountain communities to multiple risks of natural disasters. These changes have stressed the local production system and increased the trends of outmigration of rural male youths in the region.

However, the increasing trends of male outmigration not only provided stability to rural economy in terms of remittances, but also marginally improved women's access to education, development opportunities, leadership, decision-making power, natural resource management and growing market from local to global levels. These changes are contributing towards social, economic and political empowerment of rural women and providing them opportunities to be involved in decision-making processes from family to village levels. However on the other hand, increasing male outmigration has further worsened the socio-cultural life of women through increasing their hardships, drudgeries and marginalisation in the Himalaya. As a result, the workload of agricultural and livestock sectors on women has increased considerably reducing drastically the time available for their personal and child care. Furthermore, the women still constitute the highest proportion of rural population affected by natural disasters primarily due to a lack of preparedness, information and exposure. Hence, it is highly imperative to improve rural livelihood and create opportunities for employment both in traditional and in non-traditional sectors for both men and women in rural areas.

Recent studies including the present research indicate that women's access to education is improving marginally, and they are also participating in local development opportunities, leadership, decision-making process and in natural resource governance. This shows that educated women in the region have been taking advantage of the increased flow of resources and the resultant opportunities of socio-economic development for their empowerment and sustainability. As a result, mountain women who suffered from marginalisation and underdevelopment for thousands of years have been able to attract the attention of national as well as international agencies for their sustainable development. Now it is the responsibility of government and policy makers to identify, build and improve their capabilities to take up the challenges in order to benefit from these opportunities of global change. These good examples and practices, if extended in other areas through appropriate planning, go a long way in women's empowerment and mainstreaming all across the Himalayan Mountains (Banerjee et al., 2011; Dang et al., 2009; Nautiyal, 2003). This underlines the need of further research to evolve best possible strategies for sustainable mountain development in the future, particularly in emerging sectors of rural livelihood in Himalaya, such as forest conservation, village-based eco-tourism, dairying and horticulture and value chain development.

Notes

1 This chapter was originally published as Prakash C. Tiwari and Bhagwati Joshi, Gender processes in rural out-migration and socio-economic development in the Himalaya. *Migration and Development*, 5(2): 330–350, 2016, available online: www.tandfonline.com/doi/full/10.1080/21632324.2015.1022970

2 Both Village Panchayats (Village Councils) and Van (meaning forest) Panchayats (Village Forest Councils) are democratic and constitutional institutions. Village Panchayats are created for the decentralisation of developmental governance, whereas Van Panchayats are created only for the participatory management of village forests. In every village there has to be Village Panchayat, but for constituting Van Panchayat the Village Panchayat needs to move a proposal to civil administration for this purpose. This means that while in one village there may be both institutions, in another village there may only be a Village Panchayat.

References

Agarwal, B. 2010. 'Does women's proportional strength affect their participation? Governing local forests in South Asia.' *World Development*, 38(1): 98–112.

Agarwal, B. 2009. 'Gender and forest conservation: The impact of women's participation in community forest governance.' *Ecological Economics*, 68: 2785–2799.

Aguilar, L. 2006. Climate Change and Disaster Mitigation: Gender Makes the Difference, World Conservation Union (IUCN).

Ashish, M. 1983. 'Agricultural economy of Kumaon Hills: A threat to ecological disaster.' In Singh, O. P. (ed.), *The Himalaya: Nature, Man and Culture*. New Delhi: Rajesh Publications, pp. 233–245.

Banerjee, S., Gerlitz, J. Y. and Hoermann, B. 2011. Labour Migration as a Response Strategy to Water Hazards in the Hindu Kush Himalayas. Nepal: International Centre for Integrated Mountain Development.

Baniya, B. K., Tiwari, R. K., Chaudhary, P., Shrestha, S. K. and Tiwari, P. R. 2005. 'Planting materials seed systems of finger millet, rice and taro in Jumla, Kaski and Bara Districts of Nepal.' *Nepal Agriculture Research Journal*, 6: 39–48.

Beniston, M. 2003. 'Climatic change in mountain regions: A review of possible impacts.' *Climatic Change*, 59: 5–31.

Brody, A., Demtriades, J. and Esplen, E. 2008. *Gender and Climate Change: Mapping the Linkages: A Scoping Study on Knowledge and Gaps*. Susses, Bridge: University of Sussex.

Byers, E. and Sainju, M. 1994. 'Mountain ecosystem and women: Opportunities for sustainable development and conservation.' *Mountain Research and Development*, 14(3): 213–228.

CIDA. 2002. *Gender Equality and Climate Change: Why Consider Gender Equality When Taking Action on Climate Change?* Hull: Canadian International Development Agency (CIDA).

Clancy, J. and Skutsch, M. 2003. The Gender – Energy – Poverty Nexus, DFID, www.sparknet.info/goto.php/view/21/file.htm

Dang, J. P., Malhotra, K. and Ghai, D. 2009. 'Uttarakhand women entrepreneurs "the hand that rocks the cradles rules the world".' *Political Economy Journal of India*, July–Dec, 21: 100–105.

Dennison, C. 2003. From Beijing to Kyoto: Gendering the International Climate Change Negotiation Process. In 53rd Pugwash Conference on Science and World Affairs – Advancing Human Security: The Role of Technology and Politics, www.pugwash.org/reports/pac/53/dennison.htm

Department for International Development. 2007. Gender Equality Action Plan, 2007–2009, UK: DFID, http://users.ox.ac.uk/~qehwemc/documents/DFID-Gender-equality-plan-2007.pdf

Esplen, E. 2009. *Gender and Care: Overview Report*. Bridge: Development and Gender.

FAO. 2011. *The State of Food and Agriculture: Women in Agriculture: Closing the Gender Gap in Development*. Rome: Food and Agricultural Organisation (FAO).

Fisher, J. 2006. For Her It's the Big Issue: Putting Women at the Centre of Water Supply, Sanitation and Hygiene, Water Supply and Sanitation Collaborative Council, http://washcc.org/pdf/publication/FOR_HER_ITs_THE_BIG_ISSUE_Evidence_Report-en.pdf

Gupta, M. K. 2007. Promoting Self Sufficiency through Carbon Credits from Conservation and Management of Forests: Unpublished Masters Research paper, Submitted to the faculty of Clark University, Worcester, Massachusetts, in partial fulfillment of the requirements for the degree of Master of Arts in the department of International Development, Community and Environment (IDCE).

Hassan, R., Scholes, R. and Ash, N. (eds.). 2005. *Ecosystems and Human Well-Being: Current State and Trends, Volume 1: Findings of the Condition and Trends, Working Group of the Millennium Ecosystem Assessment.* Washington: Island Press.

Hoermann, B. and Kollmair, M. 2009. Labour Migration and Remittances in the Hindu Kush-Himalayan Region, ICIMOD Working Paper No. 2009. International Centre for Integrated Mountain Development, Kathmandu, Nepal.

Huddleston, B. and Ataman, E. 2003. Towards a GIS-Based Analysis of Mountain Environments and Populations, Environment and Natural Resources Working Paper No. 10. Food and Agriculture Organization of the United Nations, Rome.

ICIMOD. 2011. Labour Migration and Remittances in Uttarakhand, Case Study Report, ICIMOD, Kathmandu, Nepal.

ICIMOD. 2010. Gender Perspectives in Mountain Development: New Challenges and Innovative Approaches Periodical Publication on Sustainable Mountain Development, International Centre for Integrated Mountain Development (ICIMOD), No. 57, 2010, Kathmandu.

ICIMOD. 2009. The Changing Himalayas: Impact of Climate Change on Water Resources and Livelihoods in the Greater Himalayas, ICIMOD, Kathmandu, Nepal.

ICIMOD. 1999. Searching for Women's Voices in the Hindu Kush Himalayas, ICIMOD, Kathmandu, Nepal.

IPCC. 2014. 'Summary for policymakers.' In Field, C. B. et al. (eds.), *Climate Change 2014: Impacts, Adaptation, and Vulnerability, Part A: Global and Sectoral Aspects.* Contribution of Working Group II to the Fifth Assessment Report of the Intergovernmental Panel on Climate Change. Cambridge, UK and NewYork, NY, USA: Cambridge University Press, pp. 1–32.

IPCC. 2007. *Climate Change 2007: Impacts, Adaptation and Vulnerability.* Working Group II contribution to the Intergovernmental Panel on Climate Change Fourth Assessment Report. Cambridge: Cambridge University Press.

Jolly, S. and Reeves, H. 2005. Gender and Migration Overview Report, BRIDGE Cutting Edge Pack, UK: Institute of Development Studies (IDS), www.bridge.ids.ac.uk/reports/CEP-Mig-OR.pdf

Joshi, B. and Tiwari, P. C. 2014. Women and Sustainable Mountain Development: An Illustration of Indigenous Women's Adaptation to Climate Change in Himalaya. In Sarah, S. Aneel et al. (eds.), Proceedings of Sixteenth Sustainable Development Conference: Creating Momentum: Today is Tomorrow, Sustainable Development Policy Institute, Islamabad, Pakistan, pp. 131–144.

Kant, P. 2011. The Critical Importance of Forest Carbon Sink in the Green Economy of the Hindu Kush-Himalayan Mountain Systems, Institute of Green Economy Working Paper No. 26, New Delhi, India.

Kelkar, G. 2005. 'Development effectiveness through gender mainstreaming, gender equality and poverty reduction in Asia.' *Economic and Political Weekly*, 40: 4690–4699.

Khosla, P. and Pearl, R. 2003. Untapped Connections: Gender, Water and Poverty, New York: WEDO, New York, www.wedo.org/library.aspx?ResourceID=1

Kollmair, M., Manandhar, S., Subedi, B. and Thieme, S. 2006. New Figures for Old Stories: Migration and Remittance in Nepal, www.migrationletters.com

Kumar, A. 2013. 'Demystifying a Himalayan tragedy: Study of 2013 Uttarakhand disaster.' *Ecology*, 1(3): 106–116, July–September, 2013 (ISSN: 2321–4155).

Leduc, B. and Shrestha, A. 2008. Gender and Climate Change in the Hindu Kush-Himalayas: Nepal Case Study, International Centre for Integrated Mountain Development (ICIMOD), Kathmandu, Nepal.

Lonergan, S. 1998. The Role of Environmental Degradation in Population Displacement, Environmental Change and Security Project Report No. 4, pp. 5–15.

Macintosh, D. 2005. 'Asia, eastern, coastal ecology.' In Schwartz, M. (ed.), *Encyclopedia of Coastal Science*. Dordrecht: Springer, pp. 56–67.

Maithani, B. P. 1996. 'Towards sustainable hill area development.' *Himalaya: Man, Nature and Culture*, 16(2): 4–7.

Malhotra, A. and Schuler, S. R. 2005. Women's Empowerment as a Variable in International Development in Measuring Empowerment: Cross-Disciplinary Perspectives by Deepa Narayan-Parker, Deepa Narayan, World Bank Publications.

Malhotra, A., Schuler, S. R. and Boender, C. 2002. Measuring Women's Empowerment as a Variable in International Development. Background Paper Prepared for the World Bank Workshop on Poverty and Gender: New Perspectives.

Malla, G. 2008. Climate change and its impact on Nepalese agriculture. *The Journal of Agriculture and Environment*, 9(June 2008).

Mamgain, R. P. 2003. Out-migration among rural households in Uttaranchal: Magnitude and characteristics. *Labour and Development* (special issue on migration), 9(2): 259–287.

Masika, R. 2002. Editorial, in 'Gender, Development and Climate Change', Masika, R. (ed.), *Oxfam Gender and Development Journal*, 10(2), Oxfam, www.oxfam.org.uk/resources/downloads/FOG_Climate_15.pdf

Mehta, M. 2007. Gender and Disasters, Draft Report, International Centre for Integrated Mountain Development (ICIMOD), Kathmandu, Nepal.

Mishra, Y. D., Pal, Anju, Bharadwaj, N., Mishra, R. and Gulta, G. R. 2008. 'Constraints in Joint Forest Management (JFM): An experience of Kumaun in Uttarakhand (India).' *Progressive Research*, 3(1), pp. 38–42.

Mitchell, T., Tanner, T. and Lussier, K. 2007. We Know What We Need! South Asian Women Speak Out on Climate Change Adaptation, Action Aid International, London and the Institute of Development Studies (IDS).

Mukherjee, P. 2003. Community Forest Management in India: The Van Pan-chayats of Uttaranchal. In Paper Submitted to the XII World Forestry Congress, 2003, Quebec City, Canada.

Nautiyal, A. 2003. 'Women and development in the Garhwal Himalayas.' *Asian Journal of Women's Studies*, 9(4): 93–113.

Nellemann, C., Verma, R. and Hislop, L. (eds.). 2011. *Women at the Frontline of Climate Change: Gender Risks and Hopes.* A Rapid Response Assessment. United Nations Environment Programme, GRID-Arendal. ISBN: 978-82-7701-099-1.

Oxaal, Z. and Baden, S. 1997. *Gender and Empowerment: Definitions, Approaches and Implications for Policy*, Bridge Report No. 40. Sussex: Institute of Development Studies.

Ramachandran, N. 2006. *Women and Food Security in South Asia.* London: Palgrave Macmillan/UNU-WIDER.

Rasul, G. 2014. 'Food, water, and energy security in South Asia: A nexus perspective from the Hindu Kush Himalayan region.' *Environmental Science and Policy*, 39: 35–48.

Ratha, D. and Xu, Z. 2008. *Migration and Remittances Factbook 2008.* Washington, DC: World Bank.

Rawat, J. S. 2009. 'Saving Himalayan rivers: Developing spring sanctuaries in headwater regions.' In Shah, B. L. (ed.), *Natural Resource Conservation in Uttarakhand.* Haldwani: Ankit Prakshan, pp. 41–69.

Reuveny, R. 2007. 'Climate change-induced migration and violent conflict.' *Political Geography*, 26: 656–673, Elsevier.

Sah, M. P. and Bartarya, S. K. 2006. 'Landslide hazards in the Himalaya: Strategy for their management.' In Valdiya, K. S. (ed.), *Coping with Natural Hazards: Indian Context.* Pune: Orient Blackswan, pp. 165–178.

Sever, C. 2005. *Gender and Water: Mainstreaming Gender Equality in Water, Hygiene and Sanitation Interventions.* Bern: SDC.

Sharma, B. and Banskota, K. 2006. Women, Water, Energy, and the Millennium Development Goals: Lessons Learned and Implication for Policy. In Renewable Energy Options in the Himalaya, ICIMOD Newsletter 49, ICIMOD, Kathmandu.

Sharma, E., Bhuchar, S., Xing, M. and Kothyar, B. 2007. 'Land use change and its impact on hydro-ecological linkages in Himalayan watersheds.' *Tropical Ecology*, 48(2): 151–161.

Sherpa, D. 2007. New Vulnerabilities for Mountain Women: A Different Light on the Greater Himalaya, Final Draft Report, International Centre for Integrated Mountain Development (ICIMOD), Kathmandu, Nepal.

Shyamsundar, P. and Ghate, R. 2011. Rights, Responsibilities and Resources: Examining Community Forestry in South Asia Working Paper No. 59(11), South Asian Network for Development and Environmental Economics (SANDEE), Kathmandu, Nepal.

Sidh, S. N. and Basu, S. 2011. 'Women's contribution to household food and economic security: A study in Garhwal Himalaya, India.' *Mountain Research and Development*, 31(2): 102–111, May.

Singh, A., Svensson, J. et al. 2010. 'Gender-disaggregated data for assessing the impact of climate change.' In Dankelman, I. (ed.), *Gender and Climate Change: An Introduction*. London: Earths.

Singh, J. S., Pandey, U. and Tiwari, A. K. 1984. 'Man and forests: A central Himalayan case study.' *Ambio*, 13: 80–87.

Singh, P. and Bengtsson, L. 2004. 'Hydrological sensitivity of a large Himalayan basin to climate change.' *Hydrological Processes*, 18: 2363–2385.

Singh, S. P. 2007. Himalayan Forest Ecosystem Services: Incorporating in National Accounting, Central Himalayan Environment Association, Nainital, Uttarakhand, India.

Singh, S. P., Singh, V. and Skutsch, M. 2010. 'Rapid warming in the Himalayas: Ecosystem responses and development options.' *Climate and Development*, 2(3): 221–232, (12).

Stromquist, N. 2002. 'Educating as a means for empowering women.' In Parpart, J. L., Rai, S. M. and Staudt, K. (eds.), *Rethinking Empowerment*. London: Routledge.

Tiwari, P. C. and Joshi, B. 2014a. 'Environmental changes and their impact on rural water, food, livelihood, and health security in Kumaon Himalaya.' *International Journal of Urban and Regional Studies on Contemporary India* (Center for Contemporary India Studies, Hiroshima University), 1(1): 1–12, 2014.

Tiwari, P. C. and Joshi, B. 2014b. 'Land use changes and their impact on water resources in Himalaya.' In Grohmann, E. et al. (eds.), *Environmental Deterioration and Human Health: Natural and Anthropogenic Determinants*. Dordrecht: Springer Science+Business Media, pp. 389–399, (2014). DOI: 10.1007/978-94-007-7890-0-18.

Tiwari, P. C. and Joshi, B. 2013. 'Changing monsoon pattern and its impact on water resources in Himalaya: Responses and adaptation.' In Palutikof, J., Boulter, S. L., Ash, A. J., Smith, M. S., Parry, M., Waschka, M. and Guitart, D. (eds.), *Climate Adaptation Futures*. UK: Wiley Publishing Company. Chapter 29, pp. 633–644, 2013.

Tiwari, P. C. and Joshi, B. 2012a. 'Environmental changes and sustainable development of water resources in the Himalayan headwaters of India.' *International Journal of Water Resource Management*, 26(4): 883–907, 2012. DOI: 10.1007/ s11269-011-9825-y.

Tiwari, P. C. and Joshi, B. 2012b. 'Natural and socio-economic factors affecting food security in the Himalayas.' *Food Security*, 4(2): 195–207. DOI: 10.1007/s12571-012-0178-z.

UNDP. 2010. Power, Voice and Rights: A Turning Point for Gender Equality in Asia and the Pacific, Asia Pacific Human Development Report, Colombo: United Nations Development Programme (UNDP) Regional Centre for Asia Pacific.

UNEP. 2004. The Fall of Water, United Nations Environment Programme, GRID-Arendal, www.grida.no/publications/fall-ofthe-water/

UNEP-WCMC. 2002. Mountain Watch: Environmental Change and Sustainable Development in Mountains, UNEP, Nairobi, www.unep-wcmc.org/mountains/mountainwatchreport/ (accessed 11 March 2012).

United Nations. 2006. Empowerment of Women: Access to Assets. In Write up of Session 4, International Forum on the Eradication of Poverty, 15–16 November 2006, United Nations, New York, New York: UN, www.un.org/ esa/socdev/poverty/PovertyForum/Documents/bg_4.html

Uniyal, A. 2013. 'Lessons from Kedarnath tragedy of Uttarakhand Himalaya, India.' *Current Science*, 105(11), 10 December 2013.

Verma, R. 2001. Gender, Land and Livelihoods: Through Farmers Eyes, International Development Research Centre (IDRC), Ottawa.

Woodroffe, C. D., Nicholls, R. J., Saito, Y., Chen, Z. and Goodbred, S. L. 2006. 'Landscape variability and the response of Asian mega deltas to environmental change.' In Harvey, N. (ed.), *Global Change and Integrated Coastal Management: The Asia-Pacific Region*. Springer, pp. 277–314.

World Bank. 2005. *Drivers of Sustainable Rural Growth and Poverty Reduction in Central America*. Washington, DC: World Bank.

World Bank. 2009. *India Receives World's Largest Remittance Flows*. Washington, DC: World Bank, http://worldbank.org/E6Y6IE0CQ0 (accessed 14 November 2012).

Yaro, J. A. 2006. 'Is deagrarianisation real? A study of livelihood activities in rural northern Ghana.' *Journal of Modern African Studies*, 44(1): 125–156.

10 Climate change, drought and vulnerability

A historical narrative approach to migration from Western Odisha, India

Architesh Panda

Impacts of climate change and related implications on population mobility have given rise to a widespread debate, research and predictions on future implications of climate-induced migration sometimes described as 'climatic turn'(Neverla 2007) in explanation of migration (Stern 2006; Panda 2010; Morrissey 2009; Foresight 2011; Piguet 2012). Internal migration has already been recognised as a common response to environmental stressors and there is broad agreement that internal migration often intensifies following major droughts or famines. Indeed, a range of studies has documented such patterns of mobility (Findley 1994; Ezra 2001; Perch-Nielsen 2004). However, in the context of climate change despite many numerical predictions on the number of people who might be displaced or have already been displaced (Myers 2002; Stern 2006; Christian Aid 2007), the empirical basis of such research remains weak. Important constraints arise due to lack of data on migration and nature of complexity mostly in developing countries, where people are highly vulnerable to the impacts of climate change. However, despite empirical and conceptual constraints, the discourse on climate-induced migration has changed in the last decade from controversial concepts such as 'environmental refugees' and 'climate refugees' (El-Hinnawi 1985) to 'environmentally-induced migration' (Hugo 2008) and 'environmentally motivated migrants' (Renaud et al. 2007) recognising the difficulty in labelling refugees in legal terms and also the recognition of multiple drivers of migration including physical vulnerability and climate change. The IPCC fifth assessment report, for example, recognised that migration can be an effective adaptation strategy in the face of climate change and further mentions about the low confidence in quantitative projections of changes in mobility, due to its complex, multi-causal nature (IPCC 2014).

The mainstream migration literature identifies many drivers of internal migration such as expected income differential between the origin and the destination, income variability, diversifying risk, education and natural disasters (Lucas 2015). However, recently climate change studies have tried to examine the role of impacts of climate variability and change as separate drivers for migration from affected areas. However, there is a clear lack of research on how climate variability and change affect income levels and variability, for example, review of the literature by Lilleør et al. (2011) shows a lack of research into how climate change or climate variability affect economic drivers of migration. This partly arises because of the fact that climate change will affect migration through its influence on economic social and political drivers and the range and complexity of the interactions between these drivers means that it will rarely be possible to distinguish individuals for whom environmental factors are the sole driver. Nonetheless, the scientific evidence points towards increasing evidence of exposure of people to natural disasters and environmental changes such as higher number of people living in floodplains in the future (Foresight 2011). However scientists often make a distinction between 'climate change' and 'climate variability'. While climate change is a shift in average rainfall and temperature in the long term, climate variability is an increase in the variance of these factors also resulting in extreme weather events (Easterling et al. 2000; Rowhani et al. 2011). Thus, climate change broadly can have two types of impacts on migration, that is, migration arising due to frequent extreme events and due to change in increase of the variance of rainfall and temperature and other climatic parameters. In both the cases, however, attributing migration directly to impacts of climate change or variability has been a complex task given that migration is a second-order impact of any climate disaster. Stott et al. (2016), for example, in a careful review of attribution studies find clear evidence of human influence having increased the probability of many extremely warm seasonal temperatures and reduced the probability of extremely cold seasonal temperatures in many parts of the world. The evidence for human influence on the probability of extreme precipitation events, droughts and storms is more mixed. The study also finds that geographical coverage of events remains dispersed. It is thus extremely difficult to find a direct link except in few cases where climate change has established to be the major factor in natural disasters and consequent displacement. Recognising the limitations of scientific and empirical research in the subject, this study focuses on slow-onset disaster and assumes that case study analysis can serve as a starting point of understanding climate-induced migration from disaster-prone areas and climate hotspots around the world.

Background

Among the many impacts of climate change, analysis of displacement of people from the slow-onset and rapid-onset disasters at a national scale in India has received increasing attention in the context of climate change and weather variability impacts (Rajan 2008; Viswanathan et al. 2015; Murali and Afifi 2014). Large populations in India live in areas likely to experience increased riparian flooding and increasing water stress as a result of climate change. A significant number will also be affected by coastal flooding, while others might have to migrate because of coastal and riverbank erosion. India is a country with various socio-climatic conditions with a large number of people depending on agriculture. For last many decades, people are migrating from rural to urban areas for livelihood and higher incomes. However, climate change might result in two types of displacement and migration in India. First, increased migration is likely within India due to the effects of climate change on slow-onset and rapid-onset disasters such as drought, desertification and sea level rise, water scarcity and low food productivity, floods, cyclones and melting glaciers. Second, climate change might lead to increased flow of migrants from neighbouring countries due to the accelerated effects of climate change.

In recent years the growing concern over climate change has ignited the debate on climate migration and its implications in India. According to the Internal Displacement Monitoring Centre around 3,655,000 people were displaced due to natural disasters in India in 2015. In 1998 described as super El Nino year, 8 million people were displaced in India due to floods across twelve northern states. Further, multiple and repeated displacements in the same parts of India point to areas of particularly high exposure and vulnerability. India experiences high levels of displacement along its east coast, where communities are exposed to tropical storms from the Bay of Bengal, and in the Ganges, Brahmaputra and Yamuna river basins in the north and northeast of the country. In September, the worst floods to hit Jammu and Kashmir in 50 years displaced around 812,000 people in urban areas of the state (IDMC 2015).

Apart from floods and cyclones slow-onset disaster such as drought is also a major reason for displacement in India and might intensify in future due to climate change and desertification. Climate change is expected to increase the severity of drought especially in Western India, and about one-fourth of the area of Gujarat and 60 per cent of the areas of Rajasthan are likely to experience acute physical water scarce conditions. The river basins of Mahi, Pennar, Sabarmati and Tapi are likely to experience constant water scarcity and shortage (NATCOM 2004).

A large part of the coastal regions of India is at risk of accelerated sea level rise, intensification of cyclones and larger storm surges. Increasing adverse effects of climate change along the Indian coasts may induce many people to migrate from the low-lying and risky areas. India was estimated to have the second-largest population located in the low-elevation coastal zone of 63 million and seventh in terms of area, that is, 82,000 sq. km. The Indian region is densely populated, stretches over 7,500 km and is inhabited by more than a 100 million people in nine coastal states (NATCOM 2004: 108). Further, the sea level rise is expected to be between 15 cm and 38 cm by the middle of this century and between 46 cm and 59 cm by the end of the century. A one-meter sea level rise is projected to displace approximately 7.1 million people in India, and about 5,764 sq. km of land area will be lost, along with 4,200 km of roads (NATCOM 2004: 114). Several cases of displacement due to climate change have been reported in recent years. For example, *The Telegraph* (2006) reported that submergence of the Lohachara Island in India's Sundarban has led people to move to the nearby Sagar Island. Recently in the Mahanadi delta, the state government of Odisha is resettling 571 families due to severe coastal erosion in the Kendrapada district. Over the years many villages have lost their land to the sea. Approximately 32 per cent of India's coastal area will be at risk of inundation with sea level rise and intensified storm surges along with an additional 7,640,416 people at risk of storm surge and sea level rise.

Despite the overall evidence of impacts of climate change on migration, there have been very few studies in India conducted at the local level to understand the impacts of climate change on migration. The study by Murali et al. (2014) in Eastern India finds that migration is caused directly (shorter and delayed rain seasons) and indirectly (decreased income due to less crop production caused in turn by less rain) by water scarcity. Similarly, Deshingker (2012) examines the link between remittances and migration and finds that for internal migration, remittances are too low and mostly spent on consumption purposes. Similarly, Viswanathan and Kavi Kumar (2015) examine the impacts of a decrease in crop yields on migration. However, although these studies have provided important insights into the link between migrations, weather variability and climate change, there has been fewer attempts to analyse the linked socio-economic factors historically and the role of weather variability which triggers migration. This study makes a distinction between impacts of climate change and variability on migration. While impacts of climate change on such a scale in not available for the study region to establish relation with migration, historical climate variability and droughts serve as a channel to understand migration from the region. This study based on household surveys and group discussions from a drought-prone region in western Odisha examines the link between climate variability and migration from a historical perspective using a narrative approach.

Study region and methodology

The study was carried out in 2011 in four drought-prone villages in two districts of western Odisha that is Bolangir and Nuapada, having a historical record of migration, high level of poverty and regular droughts. The study used both the 182 in-depth household surveys and focus group discussion (FGD) to examine the link between climate change and migration in the study areas. Odisha, in Eastern India is an agrarian state with agriculture and animal husbandry contributing to more than 15.6 per cent of the Gross State Domestic Product (GSDP) in 2013–2014 at 2004–2005 prices. Agriculture provides employment directly or indirectly to 61 per cent of the total workforce in 2011 (Odisha Economic Survey 2013–2014). The normal rainfall of the state is 1451.2 mm and the actual rainfall received varies across districts. Climate variability and change are likely to adversely affect the agrarian state in the coming decade as evidenced from few recent studies. An analysis of temporal variation in monthly, seasonal and annual rainfall over the state during the period 1871–2006 revealed a long-term insignificant declining trend of annual as well as monsoon (June–September) rainfall, and an increasing trend in post-monsoon season (October–November) in the state. However, the analysis also shows a decreasing trend in monthly rainfall in June, July and September, and an increasing trend in August more predominant in the last ten years (Patra et al. 2012). Further, rainfall analysis also showed an increased number of dry years compared to wet years after 1950. Similarly, analysis by Tanner et al. (2007) shows that after 1961, the rainfall patterns are below the normal, suggesting a drier spell in Odisha (Mahapatra and Mohanty 2006; Patra et al. 2012). Likely impacts of climate change in Odisha show the possibility of an increase in hydrologic extremes (Ghosh and Majumdar 2006) including increasing probability of severe and extreme droughts (Ghosh and Majumdar 2007). The current available scientific studies on climate change do not have enough to provide evidence on migration except few past and future probabilistic trends on climatic parameters such as rainfall and drought. In terms of migration, the study districts of western Odisha have a long history of internal migration. Although migration occurs in other parts of Odisha, this study focuses on two districts of western Odisha only. Estimates based on extrapolation from large household survey reveal that there are about 0.58 million incidences of seasonal migration from western Odisha and the average number of migrants in a family from western region is 1.78 (Ali and Sharma 2014).

The research on climate change migration has used a broad range of methodologies including climate change projections, econometrics analysis and risk-based examination in field studies by McLeman (2013) and Fussell et al. (2014). However, for our study region, the study used a mixed approach of household-level data analysis and historical narrative approach from FGDs to examine the role of migration in the context of climate change. Using narratives and storytelling to communicate science has often been contested as a research methodology (Katz 2013). Narratives follow a particular structure that describes the cause-and-effect relationships between events that take place over a particular time period. However, storytelling within science should not be disregarded as methodology, as narratives offer increased comprehension, interest and engagement (Dahlstorm 2014). Qualitative methods such as of this study provide better insights into social construction of facts such as environmental degradation and decision making. Earlier studies have provided important insights into migration through qualitative analysis (Carr 2005; Pedersen 1995; McLeman and Smit 2006). The study by McLeman and Smit (2006) based on historical records has analysed how capital endowment played a role in migration in eastern Oklahoma in the droughts of 1930s. Similarly, this study uses a historical narrative approach to understand the history of migration in the region and its link to climate variability, drought and socioeconomic factors in the last six decades prior to the survey. Due to lack of historical data and documents on migration from the region, the study relies on household surveys and historical narratives provided on FGDs for analysis. The following sections provide an overview of household characteristics and also describe the results of historical narrative approach.

Results from household-level data

The study used household-level data to analyse different dimensions of migration in the study region such as frequency, reason and timing and remittances from migration. Migration from the study districts is already documented and widely debated in earlier studies. According to Wandschneider and Mishra (2003), for example, during the 2001 drought 60,000 people migrated from Bolangir district in Odisha. The study districts are characterised by high level of poverty, regular droughts and low level of economic development. For households in the study districts drought is an important negative shock that undermines livelihoods and well-being despite the use of various coping strategies. Earlier studies, for example, in Ethiopia have provided robust evidence that drought has important consequences for population mobility. The study support the common observation that mobility serves as a key coping strategy following drought, as well as the frequent assumption that the poor are most vulnerable to these effects (Gray and Mueller 2012). Study by Beegle et al. (2011) finds that rainfall shocks increase the probability of people leaving the village in Tanzania.

The study region is one of the high poverty regions in India and therefore, it is necessary to examine the income profile of the sample households of our study.[1] Income levels of the sample households and the proportion of people living below the official poverty line of the Government of India, as reported by households themselves, is presented in Table 10.1. It is found that almost 60.4 per cent of the total sample households considered by the study are living below the poverty line. While the percentage of people living below the poverty line is the highest at 31.9 among the marginal farmers, the percentage of larger farmers coming under below poverty line (BPL) category is the lowest at 1.6. Agriculture is the primary source

Table 10.1 Average yearly income by different farm size groups (in INR)

Categories of farmers	Average agriculture income	Average income from secondary income sources	Average remittances from migration	Average total income	Percentage of BPL households	Families having migrated members
Marginal	8,978	6,412	4,213	19,603	58 (31.9)	21 (28)
Small	22,019	6,148	3,764	31,931	42 (23.1)	22 (32.4)
Medium	55,062	12,107	678	67,847	7 (3.8)	8 (28.6)
Large	114,909	25,418	0	140,327	3 (1.6)	1 (9.1)
Total	27,342	8,338	3,247	38,927	110 (60.4)	52 (28.6)

Source: Primary Survey

Note: Figures in the parentheses indicate the percentages in each category of farmers.

of income for 82.4 per cent of the total sample households and 7.7 per cent are employed as agricultural labour. The average income of sample households generated by farming as primary sources and other secondary sources is INR 27,342 and INR 8,338 per annum, respectively. Table 10.1 also shows details of the average primary and secondary sources of incomes generated by all the categories of farmers. The average agricultural income in respect of large farmers is INR 114,909 per annum, and for marginal farmers, it is as low as INR 8,978 per annum, which indicates the staggering income differences between large and marginal farmers.

Migration is one of the important characteristics of the farmers in our study region. A large number of people migrate each year to urban areas within the state and also to the neighbouring states in search of employment. The role of remittances has already been recognised as an important component and reason for migration among farming households. Stark and Bloom (1985) argue that farming households participate in migration as a strategy to overcome constraints on production and investment resulting from missing or incomplete credit and insurance markets in rural areas. The remittances of different categories of farmers from migration are also provided in Table 10.2. As can be seen from the table, 28 per cent of the total sample households reported that one or more of their family members had migrated during the year previous to our survey. It is found that the rate of migration was higher among the small or marginal farmer households, but lowest among the large farmer households. This indicates that migration is more prevalent among small and marginal farmers who are relatively poorer and are unable to cope with regular droughts and associated climate risks. Remittances are one of the important sources

Table 10.2 Migration among different farm size groups

Categories of farmers	Yes	No	Total
Marginal	21 (28)	54 (72)	75
Small	22 (32.4)	46 (67.6)	68
Medium	8 (28.6)	20 (71.4)	28
Large	1 (9.1)	10 (90.9)	11
Total	52 (28.6)	130 (71.4)	182 (100)

Source: Primary Survey

Note: Figures in the parentheses are percentages of the total number of farmers in each category.

of income for the marginal and small farmer households. Information of the remittance from expatriates in respect of small and marginal farmers in our study region is provided in Table 10.2. It can be noticed from the table that the average remittances from expatriates are the highest in respect of marginal farmers at INR 4,213 per annum followed by INR 3,764 per annum in respect of small farmers. The combined average total income from all the three sources in respect of the sample households is presented in Table 10.2. The average total income of marginal farmers is INR 19,603 per annum and that of large farmers is INR 140,327 per annum. The large differences in income between marginal and large farmers show the existing income inequality in the study region.

When asked about the destination areas of migrants, many farmers reported that most of the people are migrating to the nearby Andhra Pradesh to work in brick kilns and few of them also migrate to urban areas within Orissa such as the capital city of Bhubaneswar to work as labourers and other petty jobs. The frequency and reasons cited for migration by the sample farmers' families are presented in Table 10.3. It can be noticed from the table that 48.7 per cent of the migrant households have reported at least one member having migrated, whereas 30.7 per cent of the households reported two members having migrated in search for employment elsewhere. In terms of the frequency of migration 67.3 per cent of the migrant families migrated only once in a year, and 17.3 per cent migrated twice in a year. It can be noticed from the table that around 97 per cent of the migrants were doing unskilled non-agricultural work in the new destination area. Another notable observation is that most household members migrated either to complement their low level of income or during crop failures from droughts. It can be noticed from the table that 65.38 per cent of migrants cited inadequate income as the main reason for migration, and 28.84 per cent cited drought as the main reason for migration.

In drought-prone regions such as ours the study tries to identify different adaptation mechanisms among households to deal with droughts. The results are presented in Figure 10.1. It shows that there are not many options for households in the region to cope with climatic risks. The most important responses were additional labour work, selling livestock and physical assets and migration. The table also shows that while migration is an important coping mechanisms among small and marginal farmers, it is absent in the case of large farmers among our sample households. It can be related to the modified migration model from study by McLeman and Smit (2006) that capital endowment can

Table 10.3 Frequency and reasons for migration among sample households

No. of persons migrated	Frequency of migration among families	Number of times migrated in one year	Frequency of number of times migrated	Types of jobs	Frequency of types of jobs	Reason for migration	Frequency of reasons for migration
1	25 (48.7)	1	35 (67.3)	Unskilled non agriculture	50 (96.15)	Drought	15 (28.84)
2	16 (30.7)	2	9 (17.3)	Service	2 (3.84)	Compensate low income	34 (65.38)
3	7 (13.46)	3	5 (9.61)	–	–	Lack of job	3 (5.76)
4	3 (5.76)	4	2 (3.84)	–	–	–	–
5	1 (1.92)	5	1 (1.92)	–	–	–	–
Total	52	Total	52	Total	52	Total	52

Source: Primary Survey

Notes: Figures in the parentheses are percentages of the total migrant households.

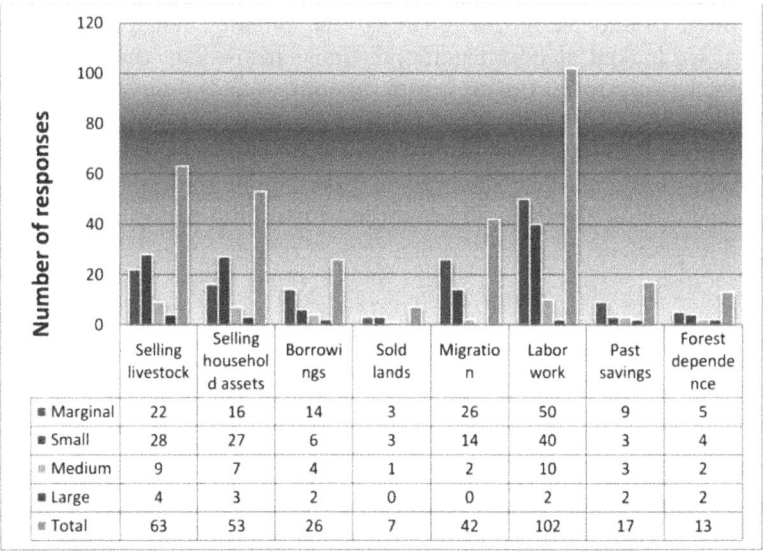

Figure 10.1 Adaptation to drought

Source: Primary survey with households

Note: Multiple Response

distinguish migration outcomes. For example land-owning households may have the option of migrating elsewhere, but might be less inclined or have less need to do so than the other.

Analysis of household-level data reveals three important aspects in our study region: (1) Small and marginal farmers have a major share in migration as compared to large farmers, (2) Although drought is a reason for migration it is not the only reason at household level to migrate and (3) remittances are not enough at household level to have any positive effect enhancing their status of poverty. Earlier research in Jharkhand by Deshingkar (2012) shows that only those farmers with larger plots of land were able to invest remittances in agriculture. Although research at household level came out with broad nature of pattern and reasons for migration in the study region, it was well enough to draw any conclusion on the links with climate variability or change. To further investigate into the reasons for migration from a drought-prone region such as ours the study employed a historical narrative approach from FGDs among farmers to examine the historical context of impacts of slow-onset disaster on migration.

Building a historical narrative

In order to understand the links between climate change and migration, we looked at how traditional oral narratives in the field area might be used to inform strategies. Interviews and observations were made in the FGDs among elderly persons in the four villages. Interviews involved questions, not asked directly but through topics, concerning oral traditions referring to past climate and environmental changes, prompted through memories of natural disasters and associated impacts.

Results

Migration was not a regular phenomenon in the study districts in western Odisha in the beginning of the 20th century and just after the independence of India. Migration was confined to tea, coffee and rubber plantations in the beginning of the century. Earlier studies have shown that poor households participate extensively in migration (Connell et al. 1976). In the post-independent period, urban pockets in India attracted labour migration from states in Eastern India such as Bihar and Odisha and Uttar Pradesh. These states in Eastern India are famous for their high level of poverty with low physical assets and a high percentage of poor people among the socially deprived classes. The rural poor have gradually concentrated in Eastern India and rainfed parts of Central and Western India, which continue to have low-productivity agriculture (Srivastava et al. 2003).

Most of the elderly respondents remember the outmigration in the immediate post-independence period and also noted migration to coffee and tea plantations in their earlier generation as oral narratives. However, it was mentioned that initial migration was mostly due to economic reasons to compensate for low income level from agriculture and the remittances which used to come were mostly invested in purchasing agricultural lands and building assets such as gold in the sending districts.

All members in the FGDs could tell of their personal experiences of natural disasters in the region, especially regular droughts in the region and related seasonal migration (see Figure 10.2). However, according to the participants, things have changed since the mega drought of 1965 and migration has become a regular phenomenon after that. Further discussion with the FGDs came out with the historical narratives of how various factors have shaped regular migration phenomenon from the districts. Farmers described that 1965 drought was a major blow

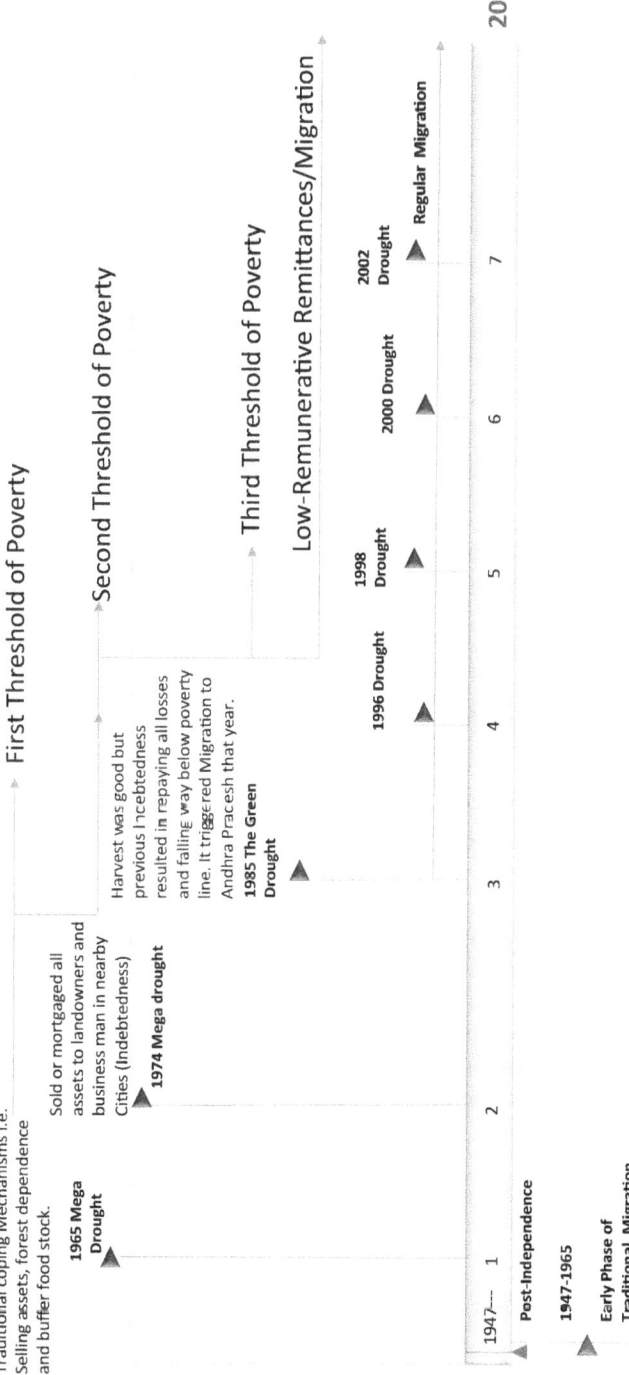

Figure 10.2 Historical narrative on migration and drought impacts

Source: Based on group discussions and FGDs in the study region

to them in the area because of droughts and they had to resort to the forest and buffer food stock at home to survive and cope with drought. Farmers mentioned that they suffered another severe drought just after a decade in 1974 which led them to sell and mortgage their assets such as land and gold resulting in high level of poverty and indebtedness among small and marginal farmers. Participants agreed to the narrative that until 1985 there was very less or non-existent migration to neighbouring states. However, that year even though the harvest and production were more than normal all marginal and small farmers had to give it back to the land owners and businessmen to repay the earlier loans. Farmers remember this year as the 'green drought' year. Farmers narrated how after that year they were left with no assets to cope with reduced level of income and how they further slid down the ladder of poverty. Since then many farmers started migrating to Andhra Pradesh and other states to work in brick kilns and it has become a regular phenomenon after that year. Historically farmers mentioned that along with reduced rainfall and regular droughts there has been a reduction of water bodies in the area leading to less income from other crops such as vegetables apart from major crop in monsoon season. The discussion revealed that although climate variability and resulting droughts have played a major role historically to induce migration, what is also worth mentioning is the role of factors such as household's assets and capital endowment in inducing migration.

Discussion

Migration as adaptation has emerged as an important research and policy area in the debate of climate change and migration nexus. The article based on household survey and historical narrative approach examined the climate change and migration nexus in a drought-prone region in Eastern Indian state of Odisha. The article makes a distinction between climate change and variability as a driver of migration from drought-prone region. The study asserts that at a local scale it is very difficult to ascertain the role of climate change in migration as it involves complex interaction of factors. While climate variability and droughts have certainly played a role in terms of increased frequency in the study districts, it cannot be concluded that climate change can be the only factor in a region to induce migration. What instead the historical narrative shows that in the case of slow-onset disasters such as drought building assets of people and adaptation measures in the sending regions can be an important strategy to make migration a more profitable adaptation and livelihood strategy. There is a large and intense debate on the links between poverty and migration and the benefits arising from migration (Connel et al. 1976; Lipton 1980; de Haan 2000). Many of the Indian scholars have emphasised the distress dimension of migration (Murthy 1991). This study finds that remuneration from migration is still minimum in the study regions and it has hardly helped the poor and marginal farmers to go out of poverty and they are trapped in the cycle of low remunerative migration and poverty. The remuneration has not been enough to build any assets at home to give sustainable earnings and break the cycle of poverty. In the study region, farmers mention about the lack of irrigation facilities and lack of renovation of traditional water bodies as an important barrier to adapting to drought in the region.

Our findings suggest three important insights into the debate on climate change and migration. First, examining migration because of slow-onset disaster such as droughts should involve more historical analysis of factors that plays a role in migration such as loss of assets and indebtedness instead of only looking at contextual factors in the current settings. Our results suggest that climate variability might impact migration via decreasing asset base and indebtedness among marginal and vulnerable farmers. Second, our study points to the fact that climate-induced migration might not be always profitable for poor and vulnerable farmers who are living in conditions of chronic poverty earnings for remittances are not enough to break the cycle of poverty. Building physical and financial assets of the poor should be a priority area for policy making. Third, it is very difficult to attribute climate change as the only cause for migration although it is playing an increasing role, rather further research can focus on how climate shocks impact household livelihood strategies, vulnerabilities and income profile for suitable intervention. Further, on the basis of conclusions, it is hard to predict how migration is likely to be affected by climate change in the future and especially in the case of slow-onset events such as droughts and sea level rise and desertification. At present there seems to be no evidence of mass migration from these regions due to climate change only although the negative link between rainfall, drought and migration seems to exist. More empirical research is certainly warranted.

Note

1 The Government of India uses Below Poverty Line (BPL) Census as an indicator of income poverty for identifying people living below the poverty line. This categorisation is based on socio-economic indicators as reflected in the quality of life of the households, and shows the number of people living below the poverty line.

References

Ali, Z. and Sharma, S. (2014). Migration trends from coastal and western Odisha: A study of migration incidence and issues. In *Studies, Stories and Canvas: Seasonal Labor Migration and Migrant Workers from Odisha*, Udaipur: Centre for Migration and Labor Solutions, Aajeevika Bureau.

Beegle, K., Joachim, De W., and Stefan, D. (2011). Migration and economic mobility in Tanzania: Evidence from a tracking survey. *The Review of Economics and Statistics*, 93(3): 1010–1033.

Carr, E. R. (2005). Placing the environment in migration: Environment, economy, and power in Ghana's central region. *Environment and Planning A*, 37(5): 925–946.

Connell, J., Dasgupta, B., Laishley, R., and Lipton, M. (1976). *Migration from Rural Areas: The Evidence from Village Studies*. Delhi: Oxford University Press.

Dahlstrom, F. Michael. 2014. Using Narratives and Storytelling to Communicate Science with Nonexpert Audiences. Proceedings of the National Academy of Sciences of the United States of America, 111(4): 13614–13620.

de Haan, A. (2000). *Migrants, Livelihoods, and Rights: The Relevance of Migration in Development Policies*. Social Development Working Paper 4. London: DFID.

Deshingkar, P. (2012). Environmental risk, resilience and migration: Implications for natural resource management and agriculture. Published 17 January 2012. *Environmental Research Letters*, 7(1): 1–7.

Easterling, D. R., Meehl, G. A., Camille, P., Changnon, S. A., Karl, T. R., and Mearns, L. O. (2000). Climate extremes: Observations, modeling, and impacts. *Science*, 289(5487): 2068–2074.

El-Hinnawi, E. (1985). Environmental refugees–United Nations environment programme. Nairobi.

Ezra, M. (2001). Demographic responses to environmental stress in the drought- and famine-prone areas of northern Ethiopia. *International Journal of Population Geography*, 7(4): 259–279.

Findley, S. E. (1994). Does drought increase migration? A study of migration from rural mali during the 1983–1985 drought. *International Migration Review*, 28: 539–553.

Foresight (2011). *Migration and Global Environmental Change: Final Project Report*. London: Government Office for Science.

Fussell, E., Hunter, L. M., and Gray, C. L. (2014). Measuring the environmental dimensions of human migration: The demographer's toolkit. *Global Environmental Change*, 28(1): 182–191.

Ghosh, S. and Majumdar, P. P. (2006). Future rainfall scenario over Orissa with GCM projections by statistical downscaling. *Current Science*, 19(6): 396–404.

Ghosh, S. and Majumdar, P. P. (2007). Nonparametric methods for modeling GCM and scenario uncertainty in drought assessment. *Water Resources Research*, 43: 1–19.

Gray, C. and Mueller, V. (2012). Drought and population mobility in rural Ethiopia. *World Development*, 40(1): 134–145.

Hugo, G. (2008). Migration, development and environment – International Organization for Migration, Migration Research Series 35. Geneva.

IDMC. (2015). *Global Estimates 2015: People Displaced by Disasters.* Geneva: Internal Displacement Monitoring Centre.

IPCC. (2014). *Climate Change 2014 Synthesis Report: Summary for Policy Makers.* Geneva: Intergovernmental Panel on Climate Change.

Katz, Y. (2013). Against storytelling of scientific results. *Nature Methods,* 10(11): 1045.

Lilleør, H. B. and Van den Broeck, K. (2011). Economic drivers of migration and climate change in LDCs. *Global Environmental Change,* 21: S70–S81.

Lipton, M. (1980). Migration from rural areas of poor countries: The impact on rural productivity and income distribution. *World Development,* 8(227): 1–24.

Lucas, B. E. R. (2015). Internal migration in developing economies: An overview, KNOMAD working paper series. Available at www.knomad.org/docs/working_papers/KNOMAD%20Working%20Paper%206_Lucas_Internal%20Migration.pdf

Mahapatra, M. and Mohanty, U. C. (2006). Spatio-temporal variability of summer monsoon rainfall over Orissa in relation to low pressure systems. *Journal of Earth System Science,* 115(2): 203–218.

McLeman, R. (2013). Developments in modelling of climate change-related migration. *Climate Change,* 117(3): 599–611.

McLeman, R. and Smit, B. (2006). Migration as an adaptation to climate change. *Climatic Change,* 76(1): 31–53.

Morrissey, J. (2009). Environmental Change and Forced Migration: A State of the Art Review, Refugee Studies Centre, Oxford Department of International Development, University of Oxford.

Murali, J. and Afifi, T. (2014). Rainfall variability, food security and human mobility in the Janjgir-Champa district of Chhattisgarh state, India. *Climate and Development,* 6(1): 28–37. doi: 10.1080/17565529.2013.867248

Murthy, R. V. (1991). Seasonal Labour Migration in Semi-Arid Areas: A Case Study of Palamuru Labour. MA dissertation, Department of Economics, University of Hyderabad.

Myers, N. (1993). Environmental refugees in a globally warmed world. *BioScience,* 43(37): 752 at 754.

Myers, N. (2002). Environmental refugees: A growing phenomenon of the 21st century. *Philosophical Transactions: Biological Sciences,* 357(1420): 609–613.

NATCOM (2004). India's first initial communication to UNFCC. Government of India, India.

Neverla, I. (2007). The climatic turn: How and why journalism has discovered climate change. Unpublished paper presented at the Conference "Communicating Climate Change" at University of Braga, Portugal, November 2007. Braga, Portugal.

Odisha Economic Survey 2013–2014. (2014). Planning and coordination department, directorate of economics and statistics. Government of Odisha, Bhubaneswar.

Panda, A. (2010). Climate refugees implications for India. *Economic and Political Weekly*, 45(20), May.

Patra, J., Mishra, A., Singh, R., and Raghuwanshi, N. S. (2012). Detecting rainfall trends in twentieth century (1871–2006) over Odisha State, India. *Climate Change*, 111: 801–817. doi: 10.1007/s10584-011-0215-5

Pedersen, J. (1995). Drought, migration and population growth in the Sahel: The case of the Malian Gourma: 1900–1991. *Population Studies*, 49(1): 111–126.

Perch-Nielsen, S. (2004). Understanding the Effect of Climate Change on Human Migration: The Contribution of Mathematical and Conceptual Models. M.Sc. Environmental Physics edn, Zurich: Department of Environmental Studies, Swiss Federal Institute of Technology.

Piguet, E. (2012). From "primitive migration" to "climate refugees": The curious fate of the natural environment in migration studies. *Annals of the Association of American Geographers*, 103(1).

Rajan, C. S. (2008). *Blue Alert*. Madras: IIT, Greenpeace.

Rao, G. B. (2001). *Household Coping/Survival Strategies in Drought-Prone Regions: A Case Study of Anantapur District, Andhra Pradesh, India*. New Delhi: Society for Promotion of Wastelands Development.

Rao, L. S. (2004). *Rural Labour Migration in Andhra Pradesh, India: Towards Addressing New Policy Spaces*. Hyderabad: Foundation for Rural and Social Development.

Rao, U. (1994). *Palamoor Labour: A Study of Migrant Labour in Mahabubnagar District*. Hyderabad: C. D. Deshmukh Impact Centre, Council for Social Development.

Renaud, F., Bogardi, J., Dun, O., and Warner, K. (2007). Control, adapt or flee: How to face environmental migration? United Nations University, Institute for Environment and Human Security (UNU-EHS), InterSections 5. Bonn.

Rowhani, P., Lobell, D. B., Linderman, M., and Navin, R. (2011). Climate variability and crop production in Tanzania. *Agricultural and Forest Meteorology*, 151: 449–460.

Srivastava, S. C., Sen, C. and Reddy, A. R. (2003), An Analysis of Growth of Pulses in Eastern Uttar Pradesh. *Agricultural Situation in India*, LIX(12): 771–775.

Stark, O. and Bloom, D. E. (1985). The new economics of labor migration. *American Economic Review*, 75: 173–178.

Stern, N. (2006). *The Stern Review on the Economics of Climate Change*. Cambridge: Cambridge University Press.

Stott, P. A., Christidis, N., Otto, F. E. L., Sun, Y., Vanderlinden, J. P., Oldenborgh, G., Vautard, R., Storch, H., Walton, P., Yiou, P. and F. W. Zwiers (2016). Attribution of extreme weather and climate-related events. *Wires Climate Change*, 7: 23–41. doi: 10.1002/wcc.380

Tanner, T. M., Nair, S., Bhattacharjya, S., Srivastava, S. K., Sarthi, P. P., Sehgal, M., and Kull, D. (2007). *ORCHID: Climate Risk Screening in DFID India, Research Report*. Brighton: Institute of Development Studies.

Viswanathan, B. and Kavi Kumar, K. S. (2015). Weather, agriculture and rural migration: Evidence from state and district level migration in India. *Environment and Development Economics*, 20: 469–492. Cambridge University Press. doi: 10.1017/S1355770X1500008X

Wandschneider, T. and Mishra, P. (2003). *The Role of Small Rural Towns in Bolangir District, India: A Village-Level Perspective, Report 2750, DFID*, World Bank Collaborative Research Project on the Rural Non-farm Economy and Livelihood Enhancement. Chatham: NRI.

11 Dynamics of distress seasonal migration

A study of a drought-prone Mahabubnagar district in Telangana

Vijay Korra

In spite of the rapid and lofty economic growth in recent decades, seasonal economic labour migration in India is mounting without falling and seems to persevere at the same pace from economically backward regions to prosperous urban and/or rural areas. Governments in India, both federal and state, focus more on development of urban centres while the countryside gets neglected in planning and budget allocation. On the other hand, more than 65 per cent of rural populace depend on agriculture, animal husbandry and forest-based activities for their living and employment purposes which is characterised by means of uncertainty, frequent natural shocks, high input cost and inadequate yields. The not so impressive atmosphere in the agriculture sector has indeed augmented labour migration which is chiefly seasonal in nature (Reddy, 2010). Seasonal exodus is either by cultivators or by daily-wage labourers or most times by both and occurs during the lean agricultural season (Deshingkar and Aktar, 2009; Vanwey, 2003). Rural mass opt migration during the redundancy periods and more often than not just for one time, but year after year for multiple purposes (Keshri and Bhagat, 2012). For the most part, such migrant forces are the landless wage seekers, small and marginal farmers chiefly from the Scheduled Castes (SCs) and Scheduled Tribes (STs) who are socio-economically and educationally bottom placed in the Indian society (Smita, 2007).

It is apparent that migratory flows have influenced the basic structure of the labour market and the way in which it functions, specifically in the informal sector, both at origin and destinations although with varied magnitude. To highlight the fact, globalisation and free and boundaryless market players have contributed very much to such alterations (Bhagat, 2011). As a result, the whole spectrum of labour migration distorted from bonded, contract and piece-rate wage mode of migration to more of individual-centric and contract-free labour movements

(Reddy, 2010; Portes, 2010). This in turn brought changes into the nature, magnitude, pattern, trends and portrayal of seasonal migration (Korra, 2011). At this juncture, questions arise such as in the current scenario, what causes labour forces to move out especially seasonally and what are their socio-economic and other associated features? How is seasonal migration distinct from other sorts of migration? How is rural distress linked to seasonal labour migration? How do the magnitude, pattern and features of seasonal migration change over the years? Does seasonal migration vary within a specified district/region, and if it does, why and how it differs within the district and between the households? These questions turn out to be extremely decisive as literature on seasonal migration talks much about migration between the districts/regions, states and countries but nothing of that sort within a specified district/region. This is true when a region predominantly represents more of distress seasonal migration than other types of migrations.

Vanwey (2003) argues that seasonal migration would depend on household land/asset ownership and income earning level on the one hand, and the profitability of agriculture and availability of wage employment in the origin on the other (Sharma, 2006). Factors such as high input cost, lack of minimum support price (MSP) for agriculture produces, inadequate credit, indebtedness, land alienation and frequent droughts are to be blamed for today's distress in Indian agriculture sector. As a result, cultivation has become impracticable for generating adequate income and employment. This constricts livelihood options and creates joblessness in rural areas (Hugo, 2009; Mamgain, 2004). This has implications on labour market and migration flows.

All these contributed to poverty and today, rural India is saddled with more than 21 per cent of 'Below Poverty Line' (BPL) populace who are mostly the landless poor and weaker sections with no or scarce household assets and resources. These are the sections that face the most economic hardships and therefore suffer from starvation, malnourishment and health hazards (Smita, 2007; Bedford et al., 2009). The vulnerable migrant workers meet distress and frantic situations time and again due to irregular monsoon wherein they don't even know how to respond to it, and thus have no choice of preferred work, destination, time of migration or duration of stay (Sharma, 2006; Smita, 2007). The necessity to ease the burden of debt induces small and marginal farmers to migrate in search of gainful employment. This has been highlighted in multiple microstudies in the past (de Haan, 2007). Paradoxically, 'somewhat better-off farmers' migrate with the intention of earning higher wage incomes so as to improve their economic and social status in the village of origin. Seasonal migrants toil in harsh conditions and go through long working hours with almost zero protection and safety at the work place. They encounter lack of shelter, sanitation, low wage payments, long working hours, lack of rest and various verbal and physical abuses. Thus, migrants are exposed to multiple risks and exploitation at the destinations (Bhagat, 2015). Moreover, there has been not much attention given to the intra-dynamics of labour movement within a particular district or region in migration studies, which are important in explaining whether the migration helps them to improve their economic conditions and welfare.

It is in this context this paper explores the dynamics of distress seasonal labour migration. Besides, it examines the employment and wage patterns of migrant workers at the destination. Also, it studies the implications of migrant earnings on their well-being. First, all this is examined in the context of Mahabubnagar district of Telangana State. Second, households from different villages are examined to highlight the importance of intra-dynamic factors which play a key role in distress seasonal migration. This aspect becomes important in the present analysis, which is not adequately explored thus far in the field of migration studies. The study analysis is based on a field survey which was conducted in 2010 in three randomly selected villages from three geographically different divisions/taluks in Mahabubnagar district of Telangana State. The study applied systematic random sampling technique in selecting the sample households. The quantitative data is supported by qualitative information that was collected along with the quantitative household survey. The present paper comprises of seven sections, including the introduction. The second section talks of magnitude of distress seasonal labour migration. The third section deals with the determinants, characteristics and pattern of distress seasonal labour migration. The fourth section is about patterns of employment and wages of distress seasonal labour migrants. The fifth section is concerned with the working and living conditions of distress seasonal migrants. The sixth section examines the patterns of remittances and income spending of seasonal migrants. The final section is the conclusion and policy implications.

Distress seasonal labour migration

To give an idea about the study area, that is, Mahabubnagar district of Telangana State, it is one of the most economically backward districts and it is the major labour supplier not only in the state but in the country as well. The labour forces from the district are recognised for their tireless and sincere hard work and known all over the country by the epithet of 'Palamur Labours'. At present, the district is witnessing high levels of labour migration particularly during the post-harvest agricultural season

Table 11.1 Sex-wise distribution of individual migrants

Akkaram		Chityala		Pata Kodangal		Grand total	
Male	Female	Male	Female	Male	Female	Male	Female
48 (53)	42 (47)	40 (60)	27 (40)	32 (53)	29 (47)	120 (55)	98 (45)

Source: Field Survey, 2010

Note: Parentheses indicate their respective percentages.

(Korra, 2011). Given the setting, this section attempts to examine the magnitude and significance of seasonal labour migration in the randomly selected study villages in Mahabubnagar district of Telangana State. On the whole, the study divulged that out of the total population (1,004), 218 individuals or 22 per cent were migrants. Of them, 55 per cent were males and the rest females. However, it differs across the villages wherein Akkaram village witnessed massive migration to other regions, with 53 per cent male and 47 per cent female migrants, followed by Chityala and Pata Kodangal villages with male migrants outnumbering female migrants (Table 11.1). Male members of a household are more inclined to migrate than their female members. Our field experience suggests that the day-to-day needs, willingness and flexibility to migrate-out any number of times to both nearby and far off places play a critical role in their migration process and male dominance is an example of that. In the case of female members, it could be argued that societal norms and reluctance due to family responsibility restrict them either directly or indirectly.

There is a positive association between the respondent's marriage status and migration. For instance, in the study region, most of the migrants were married (62%), 36 per cent were unmarried and 1 per cent were widows/widowers. Here too, male migrants leading their female counterparts. The number of unmarried male migrants outnumbered their married counterparts in Akkaram village. On the contrary, married women were larger in number than their male counterparts (table not given). It is, time and again, ascertained that temporary seasonal migration by and large is undertaken by unmarried youth and prime working-age groups (Cassarino, 2008). Further, a large section of migrant working class belong to the Lambada Tribes (STs) followed by Other Backward Classes (OBCs) and Schedule Castes (SCs) which accounted for 69 per cent, 16per cent and 15 per cent, respectively. But the same differs across the villages, where STs primarily migrated from Akkaram and Pata Kodangal villages, while OBCs and SCs from Chityala. The caste composition does not present a different portrayal as regards to gender, wherein males outnumbered female migrants (Table 11.2). STs are the most vulnerable and economically backward among the downtrodden communities.

Table 11.2 Distribution of migrants according to sub-caste and sex

Name of the castes	Akkaram		Chityala		Pata Kodangal		Grand total	
	Male	Female	Male	Female	Male	Female	Male	Female
Boya	–	–	6 (60)	4 (40)	–	–	6 (60)	4 (40)
Golla	–	–	4 (67)	2 (33)	2 (40)	3 (60)	6 (55)	5 (45)
Kammari	–	–	3 (50)	3 (50)	–	–	3 (50)	3 (50)
Kummari	–	–	3 (60)	2 (40)	–	–	3 (60)	2 (40)
Lambadas	41 (53)	36 (47)	13 (62)	8 (38)	28 (54)	24 (46)	82 (55)	68 (45)
Madiga	7 (54)	6 (46)	11 (58)	8 (42)	–	–	18 (56)	14 (44)
Telugollu	–	–	–	–	2 (50)	2 (50)	2 (50)	2 (50)
Total	48 (53)	42 (47)	40 (60)	27 (40)	32 (53)	29 (47)	120 (55)	98 (45)

Source: Field Survey, 2010

Note: Parentheses indicate respective percentages.

The other trait which is associated with seasonal migration at least in the present study region is that majority of the migrants (31%) travelled with four of their family members, 30 per cent and 26 per cent migrated together with two and three of their family members, respectively. Incidentally, in all the categories male migrants accounted for a larger share than females. The migration process in the region is such that it differs from one village to another within the same socio-economical and geographical district. For example, from Akkaram, majority of the migrants migrated together with four and three of their family members; it was two and four from Chityala; and two and three family members from Pata Kodangal village. The preponderance of four and three person movements from the same household indicates not only the intensity of migration but also the vulnerable economic condition of their households (table not given). Most of the existing studies on seasonal migration unexplored this aspect and captured several other factors, that too between two or more distinct districts/regions/states but not within a particular district.

The field survey experience apparently suggests that downtrodden social groups are more prone to witness greater family outmigration from a household. In this respect, a majority of the households belonging to SC community recorded three and four family member migration. While STs reported to have sent two of their family members, OBCs sent out two and four of their family members. Here, there is no uniformity but differs across the villages and among social groups. It is no wonder that the subjugated communities witness movements of more number of family members to other places. Ironically, such large-scale family migration is taking place at a time when governments are carving out new schemes and spending crores of rupees for the uplift-ment of the deprived communities. Yet they remained the same with little economic improvement and hence with large migrations (table not presented). It looks as if social background and household resources play a critical role in poverty-ridden communities' lives apart from other factors like pecuniary problems and redundancy (Fargues, 2008).

The age and gender facet further reveals that an outsized number of migrants constituted in the age group of 31–40 and 11–20 years with 29 per cent and 22 per cent, respectively. In most of the age groups, except for the 11–20 group, males outnumbered female migrants. Surprisingly, children aged between 0 and 10 migrated in modest numbers. Our interviews with the heads of the surveyed households suggest that the underaged children essentially migrated either along with their parents or with other family members for either employment or babysitting at the destinations (Table 11.3). It was also reported by

Table 11.3 Distribution of migrants according to age groups and sex

Villages	Akkaram		Chityala		Pata Kodangal		Grand total	
Age groups	Male	Female	Male	Female	Male	Female	Male	Female
0–10 Years	8 (62)	5 (38)	9 (69)	4 (31)	7 (58)	5 (42)	24 (63)	14 (37)
11–20 Years	16 (57)	12 (43)	6 (46)	7 (54)	2 (25)	6 (75)	24 (49)	25 (51)
21–30 Years	5 (38)	8 (62)	11 (58)	8 (42)	8 (50)	8 (50)	24 (50)	24 (50)
31–40 Years	10 (48)	11 (52)	12 (63)	7 (37)	13 (57)	10 (43)	35 (56)	28 (44)
41–50 Years	7 (58)	5 (42)	2 (67)	1 (33)	2 (100)	–	11 (65)	6 (35)
51–60 Years	–	1 (100)	–	–	–	–	–	1 (100)
61–70 Years	2 (100)	–	–	–	–	–	2 (100)	–
Total	48 (53)	42 (47)	40 (60)	27 (40)	32 (52)	29 (48)	120 (55)	98 (45)

Source: Field Survey, 2010

Note: Parentheses indicate respective percentages.

respondents that child migrants dropout from native village school and remained out of school at the destination. Thus they are deprived from their basic right to education and in the process become handicapped in enhancing their human capabilities. This would apparently deny their future employment prospects (McKenzie and Rapoport, 2006).

Descriptions of seasonal labour migrants

When examining the migrants' occupational status, the study shows that 53 per cent of migrants are from a farming background and the rest are non-farm and agricultural daily-wage labourers. Significantly, in all the occupational categories, male migrants outnumbered their female counterparts. It is pertinent to draw attention to the fact that most of the studies on migration demonstrate that land-owning households are less likely to migrate out (Connell, 2008). And land ownership means differently for different households in different districts/ regions/states/nations because of the geological nature and size of the land. Semi-arid districts like Mahabubnagar are characterised by infertile land and majority of the households possess land at least from a small to medium size. This dry region receives errant and inadequate rainfall and hence crop output is normally marginal, erratic and most of the times insufficient. Occupational diversification in the study region is negligible and dismal. And depending only on agriculture is not sufficient to survive throughout the years and searching for alternative means of survival is on the rise in the district. Hence, farming and labour force resort to temporary and especially seasonal migration (Bedford et al., 2009).

Land ownership in India is always associated with castes and occupations (Fargues, 2008). Persons from the low stratum of society are less likely to possess land. For instance, this study exposed that 35 per cent of migrants from the SC community are in fact non-farm labourers, whereas it was 33 per cent for STs and 32 per cent for OBCs. This is true even in the category of agricultural labour households. In contrast, 88 per cent of STs are cultivators and the SCs and OBCs accounted only for 1 per cent and 11 per cent, respectively. However, this fluctuates across the study villages (Table 11.4). The inference here suggests that most of the ST and OBC migrants are better off than that of SCs.

It was found that there is an inverse relationship between lack of employment, occupational diversification and migration in the study villages. It is no wonder in highlighting the fact that majority of the migrants are illiterate while very few are with primary and secondary

Table 11.4 Classification of migrants' occupation by castes

| Occupation | Akkaram | | Chityala | | | Pata Kodangal | | Grand total | | |
Castes	SC	ST	SC	ST	OBC	ST	OBC	SC	ST	OBC
Below 14 years	–	11 (100)	8 (53)	4 (27)	3 (20)	9 (75)	3 (25)	8 (21)	24 (63)	6 (16)
Agri-labour	–	6 (100)	6 (33)	5 (28)	7 (39)	–	–	6 (25)	11 (46)	7 (29)
Cultivators	1 (2)	54 (98)	–	11 (61)	7 (39)	37 (86)	6 (14)	1 (1)	102 (88)	13 (11)
Non-farm labour	12 (67)	6 (33)	2 (13)	1 (6)	13 (81)	6 (100)	–	14 (35)	13 (33)	13 (32)
Total	13 (14)	77 (86)	16 (24)	21 (31)	30 (45)	52 (85)	9 (15)	29 (13)	150 (69)	39 (18)

Source: Field Survey, 2010

Note: Parentheses indicate respective percentages.

levels of education. As expected, more female migrants were predominantly illiterate than their male counterparts. Migrants with better qualification were almost negligible in the study region (table not given). This vindicates the argument that most of the seasonal labour force comprises of illiterates and manual workers with low level of skills (Portes, 2010).

Determinants and patterns of seasonal migration

Migration during post-harvest time from the villages is chiefly a response to the changes in the native economy which emanated from the adverse effects of agricultural sector. It is revealed that majority of the workforce migrated to other destinations with the intention of daily-wage earnings (30%). However, surprisingly a good amount of labourers travelled in an attempt to survive (28%), 26 per cent of them migrated for employment and 9 per cent owing to debt burden and 7 per cent as a result of crop failure. The one difference between survival- and employment-led migrations is that the worker migrates in the absence of food grain and struggles to save his/her life at origin and thus leaves the village to another place to find work in order to endure from hunger and starvation death. Migrants' prime intent is to earn food through selling their labour in open labour market and rest is all secondary for them. In this, migrants from ST community outnumbered their SC and OBC counterparts. This differs again within the respective community. For example, STs moved out for earnings, employment and survival purposes. Among the SCs, it was for survival and wage earnings, whereas for the OBCs it was for employment and wage earnings (Table 11.5).

'Survival migration' takes place in a tricky situation where people face food grain shortage on the one hand, and do not find employment in the origin on the other. They are thus left with no option except to leave their homes in order to survive and overcome the 'distress period' in the village. And the study defines survival seasonal labour migration as a situation where people move out of their homes when there is no wage employment available for them at the place of origin and at the same time face shortage of food grain, subsequently encountering great risk of semi-starvation and hunger. It is a very commonly faced problem and is widespread among the economically deprived and indigent families in Mahabubnagar district of Telangana State (Korra, 2011). Such labour outflow takes place in the month of December (52%) followed by in the month of November (35%). It was also found that marginal number of labourers had migrated in previous years (prior

Table 11.5 Classification of individuals' reasons for migration as per castes

Reasons	Akkaram		Chityala			Pata Kodangal		Grand total		
Caste	SC	ST	SC	ST	OBC	ST	OBC	SC	ST	OBC
Survival	5 (17)	24 (83)	8 (30)	12 (44)	7 (26)	5 (100)	–	13 (21)	41 (67)	7 (12)
Employment	3 (9)	32 (91)	–	2 (25)	6 (75)	9 (69)	4 (31)	3 (5)	43 (77)	10 (18)
Earnings	3 (21)	11 (79)	8 (32)	7 (28)	10 (40)	27 (100)	–	11 (17)	45 (68)	10 (15)
Debts	2 (25)	6 (75)	–	–	5 (100)	4 (57)	3 (43)	2 (10)	10 (50)	8 (40)
Crop failure	–	4 (100)	–	–	2 (100)	7 (78)	2 (22)	–	11 (73)	4 (27)
Total	13 (14)	77 (86)	16 (24)	21 (31)	30 (45)	52 (85)	9 (15)	29 (13)	150 (69)	39 (18)

Source: Field Survey, 2010

Note: Parentheses indicate respective percentages.

to survey year) and still continues to stay at the destinations (table not given). Interestingly, a significant portion of landless workers migrated prior to land-owning migrants.

The direction of migration from the study villages is to a great extent towards Hyderabad (33%), Mumbai (24%), Pattipadu (11%) and Ahmadabad (10%). In other words, of the total number of migrants, 62 per cent migrated to places within the Telangana State (intra-state migration) and 38 per cent migrated to places outside the state territory (inter-state migration). There is no migration that took place within the district. The major intra-state destination is Hyderabad, while for inter-state migration it is Mumbai. Rural migrants preferred not to travel outside state and thus confined only to Telangana State. This is due to the fact that most of the rural migrants are farmers and decided not to travel far off places since they have to return to the village(s) before the monsoon season begins. It is the landless labourers who took too long distance migration. A significant amount (74%) of migrants travelled towards urban towns and cities across the country while remaining 26 per cent migrated to rural areas. It was learned from the field that the choice of destination would depend not just on better opportunities but migrant's household characteristics, family size, resource holdings and needs of the day (Bedford et al., 2009).

On the other hand, in rural stream of migration, females were outnumbered while males outsized their female counterparts in urban destinations. This sort of dynamic changes is significant as migration represents diversity across the villages wherein Akkaram has witnessed both rural and urban outmigration whereas the other two villages reported only urban migration. Thus a surprising 62 per cent of migrants from Akkaram moved towards rural destinations. From Chityala, a large proportion of males migrated to Hyderabad while females outnumbered their male counterparts to long distance destinations such as Surat and Vadodara. A major section of male migrants from Pata Kodangal went to Hyderabad while female migrants travelled to Mumbai (Table 11.6). This is a somewhat novel pattern and is the result of multiple causes associated to poverty, landlessness, unemployment, agriculture distress and social backwardness.

There are three forms of labour migration which exist in the villages. As stated earlier, family migration (53%) is predominant followed by 28 per cent of group migration, which is travelled by forming a group with fellow villagers, and remaining 19 per cent migrated individually to various destinations (table not given). The study defines family migration as 'if a person migrates along with three or more family members to the same destination and work in the same sectors'. Migrants from

Table 11.6 Distribution of migrants' destinations according to sex

Destinations	Akkaram		Chityala		Pata Kodangal		Grand total	
Sex	Male	Female	Male	Female	Male	Female	Male	Female
Aakaram	3 (60)	2 (40)	–	–	–	–	3 (60)	2 (40)
Ahmedabad	–	–	13 (62)	8 (38)	–	–	13 (62)	8 (38)
Bhainsa	3 (60)	2 (40)	–	–	–	–	3 (60)	2 (40)
Guntur	3 (50)	3 (50)	–	–	–	–	3 (50)	3 (50)
Hyderabad	18 (64)	10 (36)	23 (64)	13 (36)	4 (44)	5 (56)	45 (62)	28 (38)
Mumbai	–	–	–	–	28 (54)	24 (46)	28 (54)	24 (46)
Munugode	5 (31)	11 (69)	–	–	–	–	5 (31)	11 (69)
Nakirekal	3 (100)	–	–	–	–	–	3 (100)	
Pattipadu	12 (48)	13 (52)	–	–	–	–	12 (48)	13 (52)
Puttamgandi	1 (50)	1 (50)	–	–	–	–	1 (50)	1 (50)
Surat	–	–	3 (43)	4 (57)	–	–	3 (43)	4 (57)
Vadodara	–	–	1 (33)	2 (67)	–	–	1 (33)	2 (67)
Total	48 (53)	42 (47)	40 (60)	27 (40)	32 (53)	29 (47)	120 (55)	98 (45)

Source: Field Survey, 2010

Note: Parentheses indicate respective percentages.

the ST community tend to represent predominantly in all the three categories. However, this differs between households within the respective communities. Apart from economic vulnerability, family size seems to be the major cause of family migrations among STs.

Employment and wage patterns of seasonal migrants

The main advantages of migrating to economically better-off destinations are that it can offer a variety of economic opportunities like higher wage rates, higher income, better consumption and savings than that of backward origin places (EPC, 2010). This study exposed that 33 per cent of migrants engaged in construction work, 20 per cent employed in agricultural activities, 12 per cent worked as cable and drainage digging workers and 7 per cent worked in hotels and restaurants. Significantly, 15 per cent of the migrant workers did not engage in any kind of economic labour activity and this was mainly owing to underaged migrant population (below 14 years). Preceding literature on the region highlights two or three major activities in which migrants normally engage in all the destinations they migrate to. In most of the activities, migrants from the cultivator category outnumbered non-farm labourers and agricultural labourers. Interestingly, the category of workers in shops was primarily dominated by non-farm workers (Table 11.7).

Table 11.7 Patterns of migrants' employment by their usual occupation

Type of employment	Non-workers	Agri-labourers	Cultivator	Non-farm labourers	Total
Underaged children	28 (85)	1 (3)	1 (3)	3 (9)	33 (100)
Agricultural labourers	4 (9)	3 (7)	29 (67)	7 (16)	43 (100)
Construction workers	–	11 (15)	46 (64)	15 (21)	72 (100)
Brick kiln workers	3 (27)	2 (18)	4 (36)	2 (18)	11 (100)
Poultry workers	2 (40)	–	3 (60)	–	5 (100)
Cable trench workers	1 (4)	6 (23)	16 (61)	3 (11)	26 (100)
Load & unload	–	1 (20)	2 (40)	2 (40)	5 (100)
Workers in hotel/ restaurant	–	–	10 (67)	5 (33)	15 (100)
Auto/taxi drivers	–	–	1 (100)	–	1 (100)
Workers in shops	–	–	1 (25)	3 (75)	4 (100)
Housemaids	–	–	3 (100)	–	3 (100)
Total	38 (17)	24 (11)	116 (53)	40 (18)	218 (100)

Source: Field Survey, 2010

Note: Parentheses indicate respective percentages.

It was learned from the field that despite the respondent's farming background, a majority of them (migrants) swiftly adapted to urban-oriented employments that is indeed a big shift for migrant workers. However, a small segment of farmers and agricultural labourers still preferred to work in the agriculture sector for wage earnings. And this depends on various factors which need further study. It is found that female migrants were chiefly engaged in agricultural activities and brick kilns whereas males constituted largely in construction works (table not given). Significantly, seasonal migration is considerably characterised by child labour migrants. This is in contrast to the existing studies on seasonal migration. In fact, such studies highlighted that child migrants were predominant in longer duration and long distance migrations. Our interviews with respondents inform that most of the child migrants did not go to school at the destinations and either stayed at their dwellings or helped/worked along with parents at work sites. Thus they remained out of school, not just at the destination, but also at the origin due to frequent seasonal outmigration by their parents (Smita, 2007).

The other key factor which plays a critical role in obtaining employment in migrants' lives is educational level. For instance, in the study region, 62 per cent of the migrants were illiterates, and only 24 per cent were literates with primary and lower secondary education. As a result, illiterates largely worked in construction sector and in the agricultural sector (table not given). It is to be noted that with low level of education there is less likelihood of getting better jobs and upgrading their skills. It is true that illiteracy, low level of educational attainment and lower skills are the major constraints that prevent seasonal migrants from obtaining better employment/work and enhancing their earning capacity, particularly in urban destinations (Venturini, 2008).

Wage rates of seasonal migrants

On the other hand, the study as regards wage rates at various destinations found that 17 per cent of the migrants received Rs 200 wage per day, 16 per cent of them got Rs 250 and 15 per cent received Rs 180 as a daily wage. Shockingly, 18 per cent of the migrant workers received wages between Rs 50 and 100 and 12 per cent of them obtained between Rs 110 and 150. In contrast, there were few migrant workers who obtained wages up to Rs 300 per day. The lowest wage rates prevail in rural areas while the highest is in Mumbai and Hyderabad cities. In most of the wage categories, male migrants outnumbered their female counterparts. And, in few categories, females outnumbered males.

This difference is to do with the kind of work and destinations that migrants are involved in.

The wage dynamics demonstrates that the lowest wage rates (between Rs 50 and 110) were received by migrants from Akkaram while migrants from Chityala obtained medium-level wages, that is, between Rs 110 and 200, whereas migrants from Pata Kodangal earned much higher wages – between Rs 200 and 300 (table not given). Significantly, in the rural destinations there was no wage difference between male and female migrants whereas in the urban destinations, wage discrimination is widespread. Note that, in rural areas, migrants obtained wages based on the weight of cotton/chilli they picked on a particular working day, for that they received wages between Rs 2.50 and Rs 3.00 per kilogram during the peak season and at the end of the season they used to get more, and sometimes up to Rs 3.50 per day. This depends on labour supply and extent of work. Thus, rural migrants' daily wage earnings would largely depend on their ability to pluck the cotton/chilli. On average cotton/chilli picking by a young migrant could vary between 25 and 30 kilograms per day. Thus, they earn lower wages than urban migrants.

The wage patterns of migrants according to rural and urban migration streams show that 38 per cent of rural migrants earned wage of Rs 90 followed by 25 per cent and 21 per cent of the migrants obtaining Rs 100 and Rs 110, respectively. And 4 per cent of them earned Rs 50 which is far below the current wages in the Indian countryside. On the contrary, in the urban destinations, 43 per cent of the migrants received wages between Rs 180 and 200 and 27 per cent obtained daily wages between Rs 220 and 250. The number of migrants who received daily wage up to Rs 300 (3%) was insignificant. However, 9 per cent of migrants earned wages between Rs 110 and 150 per day. This is highly inadequate in the context of increasing prices of essential goods in the country. The low wage rates are a sign of the vulnerability that coexist among poor migratory communities. However, village-level dynamics shows that migrants from the better irrigated village, that is, Pata Kodangal earned better wages while lowest wage earners were from drought-prone Akkaram village (table not given). One has to bear in mind that wage payments in kind form exist not only in the rural destinations but also in the urban areas.

The other significant facet in migrants' strategy is duration of stay at the destinations. Based on migrants' duration of stay one could define whether they are seasonal or any other type of migrants. In this context, seasonal labour migrant could be defined as 'a person who stayed less than one year at the destination either for employment or in search of employment or for means of income earning during the post-harvest agricultural season and intended to return to the origin village prior to the onset of the next agricultural season'. In the study villages, 28 per cent of migrants stayed for around four months, 18 per cent stayed for about five months and 13 per cent stayed for six months at various destinations. On the contrary, 21 per cent of migrants stayed from 7 to 12 months and 10 per cent of the migrants resided from 18 to 60 months at the destinations. The rest of them stayed less than three months.

In short, greater amount of migrants (90%) resided up to or below 12 months and only 10 per cent of them stayed from 18 to 60 months at the destination. It is noteworthy to mention that though all the rural migrants stayed less than nine months, majority of them stayed less than six months at their respective destinations. In contrast, majority of the urban migrants stayed for less than 12 months (Table 11.8). It is also observed that the cycle of seasonal labour movements takes place from the study villages every year during the lean

Table 11.8 Distribution of migrants' duration of stay at destinations

Duration of stay in months	Akkaram		Chityala	Pata Kodangal	Grand total	
	Rural	Urban	Urban	Urban	Rural	Urban
2 months	–	–	1 (100)	–	–	1 (100)
3 months	7 (47)	8 (53)	–	1 (100)	7 (44)	9 (56)
4 months	28 (88)	4 (12)	13 (100)	15 (100)	28 (47)	32 (53)
5 months	16 (89)	2 (11)	11 (100)	11 (100)	16 (40)	24 (60)
6 months	–	6 (100)	3 (100)	20 (100)	–	29 (100)
7 months	–	–	18 (100)	–	–	18 (100)
8 months	–	3 (100)	10 (100)	6 (100)	–	19 (100)
9 months	5 (100)	–	–	2 (100)	5 (71)	2 (29)
12 months	–	3 (100)	2 (100)	2 (100)	–	7 (100)
18 months	–	–	–	2 (100)	–	2 (100)
24 months	–	2 (100)	5 (100)	–	–	7 (100)
36 months	–	2 (100)	–	2 (100)	–	4 (100)
48 months	–	4 (100)	–	–	–	4 (100)
60 months	–	–	4 (100)	–	–	4 (100)
Total	56 (62)	34 (38)	67 (100)	61 (100)	56 (26)	162 (74)

Source: Field Survey, 2010

Note: Parentheses indicate respective percentages.

agricultural season. This sort of migration predominantly takes place among the farming and agricultural labour communities.

More importantly, seasonal migration is not only temporary in nature but also takes place more than one time from origin place to destinations. The same was found in the study region where 12 per cent travelled twice and 8 per cent of them migrated thrice a year. The vulnerable SC and ST groups were less likely to move more number of times than OBCs. Interestingly, in the category of three-time migration, migrants from STs outnumbered the other communities (table not presented). In this, males outnumbered their female counterparts in all the categories. The village that is agriculturally and economically backward witnesses more of multiple times of labour movements while others are less likely to do so. For instance, the proportion of more than one-time migration is largely accounted for in Akkaram. This may, however, depend on other factors that require in-depth analysis to verify. The migrants who moved greater number of times usually migrated towards Hyderabad which is the nearest destination to all the three study villages. Migrants who moved to longer distances show lesser number of movements and SCs and STs are the prevalent social groups in this category. Further, migrants who belong to small and marginal farming communities make shorter trips for employment purposes while landless migrants seem to prefer to stay longer periods instead of moving out a greater number of times from their villages. The fact that was noticed in the study villages is that the landless poor do not make more frequent trips owing to their meagre economic conditions and the lack of household resources and thus choose to stay longer at the destination. It is also true that people who migrated more than once first migrated to the rural areas then returned to the origin and then again towards urban destinations.

To mention, previous studies talked of provisions of transportation offered by employer or contractor for migrant workers to travel from origin to destinations are steadily declining if not completely vanished (Bedford et al., 2009). Emergence of such a pattern is clearly brought about by this study where 60 per cent of migrants did not get any transportation facilities to reach their destinations. However, it could change when we see according to streams in which greater number of rural migrants was provided with transportation facility (91%) whilst it was only 22 per cent for urban migration (Field survey). Seasonal migration is such that its dynamics changes when migrant's characteristics change, which in turn has a bearing on other aspects.

Migrants' conditions at the destinations

In the past, quite a number of studies talked about migrants' working and living conditions at destinations although in different region-specific contexts (Deshingkar and Akter, 2009). However, presenting the same from the current study region would give an idea about how the working and living conditions are evolving in this globalised and liberalised era particularly in the case of seasonal migrants. In this regard, the present study shows that 35 per cent of migrant workforce worked nine hours per day, 33 per cent worked ten hours and the remaining 17 per cent worked the statutory eight hours per day at the destinations. A higher number of urban migrants worked longer hours than their rural counterparts in all the above-mentioned working hours' categories. Further, rural migrants worked ten hours. This implies that the bulk of the migrant workers worked more than the statutory working hours in a day. This indication is nothing but a grave violation of labour laws and labour exploitation. Therefore, the study concludes that seasonal migrants are more prone to exploitation and apathy of the governments (Bedford et al., 2009; Smita, 2007).

In regards to the problems faced by migrants, the study reveals that, 60 per cent of migrants opined problem free while 12 per cent complained about heavy work, 10 per cent reported lack of sanitation and 7 per cent stated long working hours as the major problems they faced at the destinations. Urban destinations are more likely to give troubles to the migrant workers than that of rural destinations. This is proof that temporary migrants, in particular, seasonal labour migrants are the most vulnerable to risk, insecurity and threats both at the worksite and at the living site.

Moreover, rural migrants stayed in those accommodations provided by their employers while urban migrants stayed mostly in self-made makeshift hut/tents in slum areas. Interestingly, all the rural migrants from Akkaram were provided accommodation either by employers or by contractors, while the majority of the urban migrants stated self-made makeshift arrangements. It was the practice that migrants who are recruited either by employer or by contractors are assured of their accommodation at destination. Rural migrants resided predominantly in thatched sheds while urban migrants resided in self-made makeshift dwellings in slum areas (Field survey). Note that a large proportion of migrants had to spend a lot of time to find a proper living place. The field experience taught us that migrants from the study villages have commonalities and comparability in their way of living and lifestyle regardless of their destinations. Migrants whose employment and earning opportunities are superior to their counterparts in fact live and work in better conditions.

The return from destination places to the origin village by migrant workers makes clear that rural migrants largely returned in the month of April, whereas urban migrants returned in the month of May. Interactions with migrants advocate that since seasonal migration is closely associated to agricultural activities and season, most of the rural migrants' return depends on the same at the destination. The agricultural season at destination comes to an end during the months of April and May every year. On the other hand, urban migrants too come from farming background and hence their return depends on the onset of the monsoon or end of contract/agreement with employers/contractors. This may again depend and differ based on various household characteristics such as land ownership, assets, livestock, implements and the household's decision of whether to cultivate arable land or not in a particular year (Korra, 2011). In this respect, the study reveals that 36 per cent of the migrants returned for cultivating their own arable land, 28 per cent returned for working in the village agriculture labour market during the monsoon season and 25 per cent returned as the work/season had come to an end at the destination. Marginal portion of migrants reasoned that they returned after the end of contract and some said that they returned to attend festivals and social ceremonies in the village.

Migrants' earnings, remittances and spending

Labour migrates to other prosperous destinations for work in expectation of earning more and saving from it as much as they can (Castles, 2006). There is a positive association between migrants' earnings level, savings and remittances. This could further explicate whether or not migrants benefited by migrating to other places. In this regard, village-wise migrant earnings show that the average earning of rural migrants from Akkaram was Rs 8,815 and maximum earning was Rs 22,000 (median – Rs 8,100) whereas for urban migrants the average and maximum earnings were Rs 19,269 and Rs 38,000 (median earning – Rs 20,325), respectively. In the case of Chityala, the average earning of a migrant was Rs 20,298 and maximum earning was Rs 48,400 (median – Rs 19,550), while in Pata Kodangal, the average earning of a migrant was Rs 24,643 and the maximum earning was Rs 69,000 (median – Rs 23,750). This suggests that urban migrants' earning capacity is much superior to that of rural migrants. A migrant from a backward village is likely to earn less than a migrant from a better-off village. Our experience in this regard suggests that migrants' earning capability would normally depend on two basic factors: first, the nature of work and the type of destination. Second, the migrants' educational level and skills apart from their willingness to migrate to far off urban centres in fact determine their level of income earnings. For instance, migrants from Pata Kodangal by and large moved towards Mumbai where they have earned higher wages as well as better and/or more employment opportunities. In contrast, migrants who moved towards rural areas and medium towns earned comparatively lower than the aforementioned migrant groups. The duration of stay at destinations is of course another critical element in deciding their wage income earnings and other aspects of well-being (Bedford et al., 2009).

On the other hand, the patterns of migrants' income spending that they earned from working in the destination uncovered that a whole of 32 per cent of the migrants spent their earnings on daily food consumption, 19 per cent of them spent on repayment of old debts, 14 per cent invested in agriculture, 12 per cent used up on house construction and the remaining 9 per cent spent on health, buying livestock or implements. It was found that majority of the migrants from the ST community predominately spent their earnings on daily food consumption followed by the OBCs and SC communities. But this caste seems insignificant in this regard because economic needs and demands of these migrants differ not much from each other. Yet, it is important to note that migrants from Pata Kodangal overwhelmingly invested their earnings either on construction of own house or for house repairs/renovation and they largely happened to be STs (Table 11.9). In rest of the two villages, migrants spent most of their earnings on so-called economically unproductive purposes such as daily food consumption and repayment of old debts.

The study depicts a different picture than previous studies with reference to remittances wherein most of the migrants sent remittances to their left behind family members at the origin. Herein, 94 per cent of the migrants did not send any remittances to their families back home, while only a marginal amount of them (6%) sent remittances to family members. Majority of the remitters belong to the urban migration stream and they sent for the purpose of repayment of their old debts followed by for their children's education and health checkup for their family members. However, the extent of remittances sent by migrants varies across the villages (table not presented). The channels of remittances are not like what many other studies talked about, such as banks or post offices. However, migrants from these villages themselves carried their income earnings along with them when they returned to the villages. This means that seasonal labour migration remittances made by migrants are trivial and the amount that they may remit would be dismal. This insignificant remittance pattern could be attributed mainly to their short stay, nature of work, number of worked days, number of active earning members and family size (Bedford et al., 2009).

Table 11.9 Migrants' income spending by their castes

Spending pattern	Akkaram		Chityala			Pata Kodangal		Grand total		
Caste	SC	ST	SC	ST	OBC	ST	OBC	SC	ST	OBC
Below 14 years	1 (10)	9 (90)	5 (46)	3 (27)	3 (27)	9 (82)	2 (18)	6 (19)	21 (59)	5 (16)
Consumption	5 (16)	27 (84)	6 (21)	7 (25)	15 (54)	7 (78)	2 (22)	11 (16)	41 (59)	17 (25)
Debts	4 (17)	19 (83)	2 (18)	3 (27)	6 (55)	5 (71)	2 (29)	6 (15)	27 (66)	8 (20)
Agriculture	3 (20)	12 (80)	1 (11)	3 (33)	5 (56)	4 (67)	2 (33)	4 (13)	19 (63)	7 (23)
House built	–	1 (100)	–	2 (100)	–	23 (100)	–	–	26 (100)	–
Health	–	4 (100)	2 (50)	1 (25)	1 (25)	2 (67)	1 (33)	2 (18)	7 (64)	2 (18)
Cattle/implements	–	3 (100)	–	–	–	2 (100)	–	–	5 (100)	–
Most of the above	–	1 (100)	–	1 (100)	–	–	–	–	2 (100)	–
Others	–	1 (100)	–	1 (100)	–	–	–	–	2 (100)	–
Total	13 (14)	77 (86)	16 (24)	21 (31)	30 (45)	52 (85)	9 (15)	29 (13)	150 (69)	39 (18)

Source: Field Survey, 2010

Note: Parentheses indicate respective percentages.

In order to support quantitative information, this study has also collected certain essential qualitative information which in turn enriches the analysis. First, it was asked if they would migrate even if they get work/employment within their respective villages and 36 per cent of them expressed that they wanted to migrate even if they get employment in the villages, especially during the post-harvest agricultural season. However, 64 per cent expressed their unwillingness to migrate to other places if they got work/employment in their own villages. Migrants who preferred to move out despite the availability of work locally are primarily interested in earning more wage incomes by taking up urban-oriented jobs. The key motivation for higher earnings indeed comes from a strong desire to accumulate income for construction of their own house, repair dwellings and buy land and livestock. But remember, during the monsoon season they engage in their own cultivation and afterwards, if work was available in the village labour market, they take up employment, although irregular, and thus stay back. This strategy in fact makes possible for them to prepare their arable land for the next agricultural season. Besides, by skipping migration they could also avoid risks involved in migration.

Given this context, the study collected information on migrants' perception on whether they plan to migrate again for the next agricultural season, after returning from the destinations. There were about 70 per cent of them who expressed they would like to migrate again and only 20 per cent of them opined they would not migrate in the coming or next season. Interestingly, 10 per cent of them expressed that they do not know whether they will migrate or not (Field Survey). Most of them supposed that 'their decision to migrate or not' predominantly depends on various economic, household, crop/yield and climatic conditions at the origin. Since most of the migrant households possess arable land, the magnitude of migration from these villages fairly depends on the monsoon conditions apart from other household factors.

Concluding remarks

The main crux of the chapter was to present the current dynamics of distress seasonal labour migration from Mahabubnagar district of Telangana State. More significantly, it tried to bring out the internal dynamics of seasonal migration which exists across the villages in the district. In this, the study clearly exposed that seasonal labour migration is largely by farmers and labour households, particularly in the deprived communities, that is, SCs and STs. It was mainly due to distress and dearth of work, debt burden and for survival at the time of distress in the villages. The internal dynamics of seasonal labour migration is multifaceted and varies not just across the villages but also between the households within a village economy which played a critical role in their decision to migrate out. Seasonal migration is not drastically changing their lives, but providing a mere relief during adverse times. On the other side, poor and vulnerable migrant groups are exposed to multiple problems at destinations such as hazardous working conditions, labour exploitation, lack of shelter, absence of sanitation and lack of protection and safety. Besides, they face opposition now and then from local people where they migrate to because of their fear of loss of employment and social disorder. Incidentally, Indian labour laws/regulations do not address the issues of seasonal migrants and remain ineffective. Thus, it is very important to implement and monitor the labour laws and regulations in order to safeguard the migrant workers. Furthermore, lack of welfare measures under the Inter-State Migrant Workmen Act is palpable and the laws need to be amended in order to provide government welfare benefits to the migrant population. Also, there is a need for providing social security provisions so as to ensure free and safe migratory flows in the country. Further, migrants have to be allowed to claim the benefits from government schemes at the destinations. Unrestricted and secure labour movements are essential for the development of the country. At the same time, rural areas should be developed. Otherwise distress-led labour migration cannot be avoided.

References

Bedford, C., R. Bedford, et al. (2009) 'The Social Impacts of Short Term Migration for Employment: A Review of Recent Literature', Research Report for NZAID, Population Studies Centre, University of Waikato, Hamilton.

Bhagat, R. B. (2015) 'Urban Migration Trends, Challenges and Opportunities in India', Background Paper, World Migration Report, 2015, International Organisation for Migration (IOM).

Bhagat, R. B. (2011) 'Emerging Pattern of Urbanisation in India', *Economic and Political Weekly*, Vol. 46, No. 34, August 20.

Cassarino (2008) 'Patterns of Circular Migration in the Euro–Mediterranean Area: Implications for Policy Making', CARIM analytic and synthetic notes 2008/29, circular migration series. Florence, Robert Schuman Centre for Advanced Studies, European University Institute, Florence.

Castles, S. (2006) 'Guestworkers in Europe: A Resurrection', *International Migration Review*, Vol. 40, No. 4, pp. 741–766.

Connell, J. (2008) 'Poverty, Migration and Economic Resilience in Small Island Developing States', In L. Briguglio, G. Cordina, N. Farrugia, and C. Vigilance (eds), *Small States and the Pillars of Economic Resilience*. Valletta: University of Malta, pp. 263–288.

de Haan, Arjan and Shahin Yaqub (2007) 'Migration and Poverty: Linkages, Knowledge Gaps and Policy Implications', Revised Paper UNRISD/IOM/ IFS Workshop "Social Policy and Migration in Developing Countries", 22–23 November, Stockholm.

Deshingkar, Priya and Shaheen Akter (2009) 'Migration and Human Development in India', United Nations Development Programme, Human Development Reports, Research Paper 2009/13, April 2009.

EPC (2010) 'The Impact of Temporary and Circular Migration on Migrants, Their Families and Their Countries of Origin', 12 October 2010 Workshop Report, European Policy Centre, Bonn.

Fargues, F. (2008) 'Circular Cigration: Is It Relevant for the South and East of the Mediterranean?', CARIM analytic and synthetic notes 2008/40, Robert Schuman Centre for Advanced Studies, European University Institute, Florence.

Hugo, G. (2009) 'Circular Migration and Development: An Asia-Pacific Perspective', September 2009, Multilcultural Center, Prague. Also available at: (seen 7 February 2011) http://migrationonline.cz/e-library/?x=2198523.

Keshri, Kunal and R. B. Bhagat (2012) 'Temporary and Seasonal Migration: Regional Pattern, Characteristics and Associated Factors', *Economic and Political Weekly*, Vol. 47, No. 4, January 28.

Korra, Vijay (2011) 'Labour Migration in Mahabubnagar: Nature and Characteristics', *Economic and Political Weekly*, Vol. 46, No. 2, pp. 67–70, January 8.

Mamgain, P. Rajendra (2004) 'Employment, Migration and Livelihoods in the Hill Economy of Uttaranchal', Centre for the Study of Regional Development School of Social Sciences, Jawaharlal Nehru University, New Delhi, 110067 India.

McKenzie, D. J. and H. Rapoport (2006) 'Can Migration Reduce Educational Attainment? Evidence from Mexico', World Bank Policy Research Working Paper No. 3952, Washington.

Portes (2010) 'Migration and Social Change: Some Conceptual Reflections', *Journal of Ethnic and Migration Studies*, Vol. 36, No. 10, pp. 1537–1563.

Reddy, D. Narasimha (2010) 'Land and Labour in Andhra Pradesh', *Economic and Political Weekly (EPW)*, Vol. 45, No. 39, September 25.

Sharma, N. Alakh (2006) 'Flexibility, Employment and Labour Market Reforms in India', *Economic and Political Weekly (EPW)*, Vol. 41, No. 21, May 27.

Smita (2007) *Locked Homes Empty Schools: The Impact of Distress Seasonal Migration on the Rural Poor*, New Delhi: Zubaan Publications.

Vanwey, Leah K. (2003) 'Land Ownership as a Determinant of Temporary Migration in Nang Rong, Thailand', *European Journal of Population/ Revue Européenne de Démographie*, Vol. 19, No. 2, pp. 121–145. www. jstor.org/stable/20164221. Accessed 2 December 2009.

Venturini, A. (2008) 'Circular Migration as an Employment Strategy for Mediterranean Countries', CARIM Analytic and Synthetic Notes 2008/39. Circular Migration Series, Robert Schuman Centre for Advanced Studies, European University Institute, Florence.

12 Seasonal migration from dry climatic zone

A case of rural Maharashtra

Abdul Jaleel C. P. and
Aparajita Chattopadhyay

The nexus of environment and migration has attracted a growing amount of interest among scholars and policymakers. This paper examines one of the important coping mechanism of households against environment-related shocks in a dry climatic zone. Every year, the countryside of Maharashtra witnesses seasonal migration of several households to better-endowed areas within and outside the state. In a semi-arid topography, lack of irrigation facility makes it difficult to cultivate after the monsoon. Even though unemployment persists throughout the year, the situation exacerbates after the monsoon. In the months of October and November, thousands of poor peasant households undertake migration for working in sugar factories, brick kilns, stone quarries etc. where they remain for varying periods generally until the rain starts again in their home areas. Using the household-level primary data collected from rural areas of Beed and Solapur districts of Maharashtra, we examine the characteristics, pattern and causes of seasonal migration.

Background

Although panel data on seasonal migration in India are lacking, a growing number of micro-level studies have established that seasonal migration for employment is growing (Breman, 1985, 1996; Rao, 1994; Rogaly et al., 2001). Various micro-level studies on seasonal migration have been carried in many parts of India and found that it has long been a part of the livelihood portfolio of poor people (Rao, 1994; de Haan, 2002). Several studies conducted by Breman in Gujarat and by Rogaly and others in West Bengal analyse the causes for labourers to leave their villages. Their findings show that several influential socio-economic and cultural factors motivate labourers to depart from their usual place of residence. Deshingkar and Start

(2003) found that in Andhra Pradesh those with more animals are less likely to migrate, whereas in Madhya Pradesh, having livestock significantly increases the likelihood of migration. Deshingkar (2010) says growing level of inequality and uneven growth has resulted in the large section of the population being excluded or adversely incorporated. Many of these people belong to remote rural areas and are chronically poor. They routinely migrate for work to smooth consumption, repay debts and invest in health and agriculture. Gidwani et al. (2003) found that seasonal migration is a way of covering the income shortfall and argue that seasonal migration/labour circulation is also influenced by the desire of migrants to refashion place-based identities. Circular migration helps the people at least temporarily to undercut the undesired roles thrust upon them by history. Haberfeld et al. (1999) found that those households in the less developed regions and having larger labour supply, lower level of education, low income and less number of livestock tend to raise the probability to migrate.

The climate change–migration nexus

Environmental change is likely to affect migration flows in several ways. The increase in drought or flooding may reduce livelihoods in certain areas particularly those based on agriculture, causing residents of these areas to move elsewhere to support themselves. Climate change may reduce the ability to live a productive life in their usual place of residence and consequently people may take up migration as the best strategy for increasing their life chances. The vulnerability of the human population to climate change lies at the intersection of exposure and the availability of resources to cope with the impact of such exposure. In other words, a population's resilience in the face of climate change depends on access to human, social, political and financial capitals that allow individuals and their communities to recover from destructive episodes. A combination of exposure to natural hazards, poverty-related vulnerability and resilience (the ability to adapt) determine which sections of society are more likely to suffer adverse effects of climate change.

As per the fourth assessment by the Intergovernmental Panel on Climate Change (IPCC) in 2007, rain-fed farm regions and sub-humid and arid regions are among the most susceptible zones to environmental changes. Changes in rainfall patterns can make some areas drier and more prone to drought. The situations can destroy crops and make it even harder for people, especially those dependent on rain-fed agriculture with few resources. The relationship between climate change and migration is not a linear one, but rather more complex, unpredictable and influenced by larger social, economic and political forces that shape how societies interact with their environments. Thus, migration because of environmental change needs to be analysed in the context of three interrelated characteristics: vulnerability, resilience and adaptability. As it relates to climate change, vulnerability is a function of an individual or group's capacity to anticipate, cope with, resist and recover from adverse conditions. The degree of vulnerability also reflects resilience, meaning the ability to absorb external shocks and preserve preferred life-course options in the face of environmental change. Resilience, to a large extent, depends on access to human, social, political and financial capitals that allows individuals, households and communities to recover from disasters and adapt to permanent changes in the environment. Migration has helped humans to cope with environmental changes for centuries and will continue to be one of the ways people respond to climate change. Thus, climate change has become a new driver of human migration.

Study area

Data for this study were collected from rural areas of Beed (Marathwada region, Aurangabad Division) and Solapur districts (Pune Division) of Maharashtra. This section presents the geographical context of the area where this research was carried out. Marathwada consists of eight districts of Maharashtra: Aurangabad, Beed, Latur, Osmanabad, Parbhani, Jalna, Nanded and Hingoli. This has historically been a rain shadow region with average annual rainfall of about 700 mm, but in districts like Beed, it dips down to 600 mm. Apart from the Godavari, no major rivers originate or flow through Marathwada. The region was always backward regarding the availability of natural resources and access to capital investment. Similarly, Solapur is a drought-prone area, drained by Bhima River (a head stream of Krishna river) that remains dry except during monsoon. Table 12.1 presents some basic development indices of these two districts.

Beed and Solapur have usually been prone to droughts since historical past. Almost every 20 years, a big drought has hit the region, while smaller droughts keep occurring in the interim. Extreme weather events have become more frequent in the region. There were little rains in 2010, 2011 and 2013 and severe droughts in 2012. Narayanamoorthy and Venkatachalam (2011) explained the factors

Table 12.1 Selected development indicators of Beed and Solapur districts

Indicators	Beed	Solapur
HDI (2012)	0.700	0.718
IMR (2010)	33	23
Child sex ratio (Census 2011)	801	872
Per capita NDDP (2011–2012)	51,131	74,856
Literacy (Census 2011)	73.53	77.72
Population density (Census 2011)	242	290

Source: UNDP (2012). Maharashtra Human Development Report. Pune: Yashwantrao Chavan Academy of Development Administration & New Delhi: Sage.

that compounded the water scarcity problem of the region as the undesirable cropping pattern followed in the state for a long time (in spite of water scarcity, the area under sugarcane has been consistently increasing) coupled with the massive transfer of water from irrigation to non-irrigation purposes (domestic and industrial use). According to the report of a high-level committee on inter-regional disparities appointed by the Government of Maharashtra in the year 2013, Marathwada constitutes 31 per cent crop area of the state but it uses only 14 per cent of the state's surface water, on the other hand, Western Maharashtra has 36 per cent crop area of the state, but uses 47 per cent of the water.

In the last two decades, traditional crops of the region like groundnut, jowar and soya bean that gave sustainable food and income to the farmers with less water have largely been replaced by cash crops. With failing returns from traditional crops, distressed farmers are turning to the more lucrative but water-intensive sugarcane. Sugarcane needs more than 2,000 mm of rainfall. Since there is no rainfall, whatever little surface and groundwater is available in the region is guzzled by sugarcane-growing farmers at the cost of those who grow traditional staple crops. The impact of all these on rainfed farmers has been catastrophic. The incidence of drought remains a threat to agricultural production and to the livelihoods of people depending on agriculture because cultivation has become a gambling with the monsoon. The people live in hand-to-mouth economy such as landless labourers, and small marginal farmers who do not have enough stock to sustain in the event of a drought take up many coping strategies among which seasonal migration is a prominent one. In this context, this study examines the nature of seasonal migration happening in this region.

Objectives, data and methods

This micro-level study was carried out by covering 340 households with a population of 1,881 persons from 14 randomly selected villages in Beed and Solapur districts of Maharashtra state. This paper looks into the characteristics, patterns and causes and consequences of seasonal migration from rural Maharashtra. A seasonal migrant household is defined as 'a household having, at least, one member who has stayed away from the village for the purpose of employment for at least one month during one year preceding the survey'. It is characterised by an intention of temporary stay to avoid livelihood shocks during the dry season (October to May) of the year 2012. The field work was conducted during the months of September and November in the year 2013. A detailed questionnaire covering questions on household information, pattern and causes of migration is canvassed for data collection. Household and migration information was collected from the head of the households (if the head of the household was not a migrant, then migration information was collected from a male migrant member of the household). Information on the social costs of seasonal migration was collected from a female migrant member of the household. Since the subject of this paper cannot be completely addressed by empirical material, we have gathered narratives from migrants about their story of migration expecting that at least to some extent these qualitative narratives go more in depth into the causal mechanisms underlying distress-driven migration. Results are produced by using bivariate analysis. This paper is organised in three sections. In the first section, demographic, social and economic characteristics of seasonal migrant households and migrants are discussed. The second section then discusses the pattern of seasonal migration. Finally, the third section discusses the causes and consequences of seasonal migration.

Findings

Household characteristics of seasonal migrant households and migrants

In household characteristics, we have included the demographic, social and economic characteristics of the sample households. Within the households, there are two categories of people, namely, people who took up seasonal migration (movers) and people who remained at home (non-movers). While discussing demographic characteristics of the total population, we discuss demographic characteristics of the

Table 12.2 Demographic characteristics of the total population, seasonal migrant population (movers) and population who remained at home (non-movers)

Age group	Total population			Seasonal migrants			People who remained at home		
	Male	Female	Total	Male	Female	Total	Male	Female	Total
0–14	33.8	30.1	32.1	22	17.5	19.9	55.1	54.6	54.9
15–64	59.2	63.9	61.4	72	79.3	75.4	36.4	33.3	35.1
65+	7.0	6.0	6.5	6	3.2	4.7	8.5	12.0	10.0
Percentage	54.2	45.8	100	53.5	46.5	100	55.6	44.4	100
Samples	1,020	861	1,881	653	572	1,225	365	291	656
Mean age			25.4			28.1			20.4
Average household size			5.5			3.6			1.9

Source: Field survey by the authors.

subsets of the study population, i.e., seasonal migrants and people who remained at home.

Table 12.2 presents the demographic characteristics of the study population and its subsets. A total of 340 households were successfully interviewed, and it covered a population of 1,881. Average household size in the study population was 5.5 people/household and this became 1.9 people/household at the time of migration. The average number of seasonal migrants per household in the population was 3.6. In other words, magnitude of seasonal migration within the population for the year 2012 was 65 per cent. Among the study population, out of every 100 individuals, 65 took up seasonal migration and the remaining 35 stayed back in villages. Among the working-age population (15–59 years), 80 per cent migrated as seasonal migrants. By any reckoning, the movement of 80 per cent of the total working-age population (78 per cent of overall male and 82 per cent of overall female) away from their homes and farms for nearly half of the year is a massive event in rural life.

As regarded to the religious composition, Hindus, Buddhists and Muslims were 84, 12.6 and 3.4 per cent respectively in the sample households. Caste composition of the households suggests that majority (68 per cent) of the households belong to either Scheduled Tribe (ST) or Schedule Caste (SC). It reveals that migration from rural Maharashtra is to a large extent by socially disadvantaged communities. However, by no means, seasonal migration has been restricted only to the lower castes. The relatively large number (32 per cent of the total households) of Other Backward Castes (OBCs) and General caste households also took up seasonal migration. Overall literacy rate in the population was 61 per cent (73.3 and 46.7 per cent respectively for males and females).

Agriculture being the major economic activity for many of the rural households, the researchers have attempted to understand household's ability to participate in the agricultural activities. Along with this, household's source of income, access to social security schemes, access to basic amenities and possession of other household durables were examined. Households were asked whether they had any cultivable land. It draws that in the study population, the majority of the households (58 per cent) did not own any agricultural land, and 11 per cent owned only less than one acre. In other words, over 69 per cent of households in the study population either were landless or only had a marginal land (less than one acre) for cultivation. This means that two out of every three seasonal migrant households have to make survival through wage labour and by doing other works. Another aspect is, despite owning land, small and marginal farmers also migrated. They could not generate subsistence throughout the year because their landholding is unproductive (non-irrigated). Caste-wise differential in land distribution among the study households shows that 6.5 among every 10 SC/ST households, 3.5 among every 10 OBC households and 5.5 among every 10 General caste households were landless. Average land possessed by SC/ST households is estimated at 0.62 acres, whereas it is 2.13 acres for OBC households and 1.34 acres for General caste households.

In villages, people rely heavily on wage and cultivation for subsistence. The majority of the households earn money (source of income) either as agricultural or as non-agricultural labourers. They have to do a mix of the job to manage their survival. The majority of the households (70 per cent) survive with income from more than one source. It suggests that the people have to diversify their income source because income from a single source cannot support their survival. They switch to the next alternative whenever the source of income becomes insufficient (lean). Else, some people engage in household cultivation, and the excess labourers in households go for wage employment. From the field, the researchers observed that the landless households earn as much income as agricultural and non-agricultural labourers. A few landless households cultivate on leased land (sharecropping) and work as labourers (agricultural and non-agricultural). In the same way, the households having own land earn income from cultivation, and as agricultural and non-agricultural labourers.

In this economic realm, the researchers have tried to understand the government initiatives to provide social safety provisions such as food and employment to them. In India, Public Distribution System (PDS) is an important social security programme to provide food grains to people at a subsidised rate. As far as access to PDS for the study population is concerned, 74 out of every 100 households had ration cards. Of every 100 ration cards, 69 were Below Poverty Line (BPL). This indicates that access to PDS is not universal in the population. However, a substantial number of households (69 per cent of the households with ration card) are eligible to get food grains at a subsidised rate. The other scheme to provide employment is the Mahatma Gandhi National Rural Employment Guarantee Act (MGNREGA). The mandate of MGNREGA is to provide at least 100 days of guaranteed wage employment in a financial year to every rural household whose adult members volunteer to do unskilled manual work. The scheme is implemented to help the rural poor to strengthen their livelihood resource base. It is found from the study that only one-fifth (19 per cent) of the eligible population (18+) received employment at least one day under MGNREGA during last one year (2011–2012). This paper attempted to estimate the average wage income of a male and a female worker in the village and at the destination, and how important the seasonal migration is for managing their income shortfall. The results are presented in Table 12.3.

Table 12.3 Average working days, wage and share of income from seasonal migration by male and female seasonal migrant workers of rural Maharashtra

	Male	Female
Days stayed in the village	194	194
Days stayed at the destination	171	171
Average wage labour (days) received in the village in 2012	48.3	63.3
Average wage for work in the village (Rs)	186.8	93
Average work days at the destination	140	140
Average wage at the destination	220	218
Total income from wage in the village	9,022	5,887
Income from wage at the destination	30,800	30,520
Total annual wage income	39,822	36,407
Wage income share (in terms of %) from seasonal migration	77.3	83.8

Source: Field survey by the authors.

Table 12.3 presents the average days stayed by a migrant worker in the village and at the destination, and the income from wage labour. The migration period is a dry season in their area because of aridness, drought and shortage of water. In a year, the average days stayed by a migrant worker in the village are 194 and at the destination are 171. Women get more wage days than men (63 work days for female worker against 43 for the male worker) in their villages. Average wages estimated for a male worker and a female worker in the village are Rs 187 and Rs 93, respectively. However, average wages estimated for a male worker and a female worker at the destination are Rs 220 and Rs 218, respectively. It is found that 77 per cent of the total annual wage income of a male worker and 84 per cent of the total annual wage income of a female worker come from seasonal migration. It describes how rural households of Maharashtra derive income from various sources and how critical seasonal migration is as a component in their livelihood.

We further tried to understand the amount left with a work unit after six months of seasonal migration and work. What the researchers primarily look at is whether any money is left with the seasonal migrants after repayment of advance taken by the work unit when they were in the village before migration. Three questions were asked in this respect: did they save some money from wage at the end of migration? If saved, how much money were they left with, and finally what did they do with that amount? Results in this respect are presented in Table 12.4. When asked about the money they have saved from their last seasonal migration, 54 per cent reported that they could save some money, in which 49 per cent could only save up to Rs 5,000 and 46 per cent could save up to Rs 10,000. In other words, it indicates that only 54 per cent of the migrated households could repay the advance amount taken from the labour contractor and gained a surplus. The households who could save some money after their last migration were asked about the utilisation of their savings and 41 per cent of the households reported that they had spent their saving for consumption of food or on managing their daily living. They do not gain much disposable income and thus, whatever amount they are left with from seasonal migration is often hidden as it goes straight into household's daily consumption and paying off debts. Overall, 46 per cent of work units (a work unit consists of a male and a female worker) did not save or were left with nothing after five to six months of back-breaking labour. These work units either repaid the amount they had taken as advance or failed to repay the advance through their labour. In this situation, they are compelled to take advance from labour contractor

Table 12.4 Economic benefits of migration

	Frequency	Percentage
Work units who could save money from wage received during last migration		
Saved money	183	53.8
Not saved	157	46.2
Total	340	100
Amount saved by respondents (in Rs)		
Up to 5,000	89	48.6
6,000–10,000	85	46.4
11,000–15,000	6	3.3
More than 15,000	3	1.6
Total	183	100
Utilisation of amount saved		
Spent for household consumption	75	41
Deposited in bank	40	21.9
Maintenance of house	23	12.6
Repayment of loan/debt	18	9.8
Purchased gold/other assets/ household durable	16	8.7
Others	11	6
Total	183	100

Source: Field survey by the authors.

for next year's migration. It points to the fact that seasonal migration takes place mainly for survival and repayment of debts, and a large proportion of their earnings is utilised for day-to-day expenses.

Pattern of seasonal migration

Examination of the pattern of migration is undertaken with the purpose of understanding the nature of seasonal migration from rural Maharashtra. We have examined the time of migration (the month in which migrants leave their villages and month in which they come back), duration of migration, work at the destination, the number of workplaces visited during a single season, place of migration (name of state or districts) etc. Seasonal migration in rural Maharashtra involves labourers moving between home and sugarcane fields, brick kilns, mines and stone quarries. These works are mainly characterised by the predominance of manual processes, seasonality and contract migrant labour. During the course of the year, they change their place of residence for five to six months. This forth and back movement is a routine for many households. They leave their villages between September and November every year and come back to their villages between March and April. During the dry season of 2012, 87 per cent of the sampled households left their villages in October, and 46 per cent and 41 per cent of the households respectively came back to the villages in March and April. It shows that the movement of people from rural Maharashtra is primarily seasonal in nature. Regarding the duration of migration, 49 per cent of the households left their villages for five months, and 42 per cent of the households left for six months. Sugarcane cutting was the principal occupation for almost all (98 per cent) the migrant households and the remaining 2 per cent of the migrant households were working in brick kilns and stone quarries.

The study has found that seasonal migration from rural Maharashtra is mostly inter-state (58 per cent) in nature, almost all (99 per cent) seasonal migration is family migration, in which 88 per cent go to a single destination, and 12 per cent go to multiple destinations. After finishing the first contract, some households enter into another fresh contract for working somewhere else and thus in a single season few households visit more than one destination. 33 per cent of the households were having two migrants and 24 per cent of households were having three migrants, and 24 per cent of the households had migrated entirely in the year 2012. The migration history of the households regarding the frequency of migration taken up during the last five years was also collected. During the last five years (from 2008 to 2012), 75 per cent of the sampled households have continuously (five times) taken up seasonal migration. This clearly indicates the routine nature of seasonal migration from rural Maharashtra. It is observed that, although the pattern of migration is broadly similar across the villages, the extent of migration varies among villages. There are villages where the majority of the households migrate, whereas in some villages only a few households migrate.

Causes of seasonal migration from rural Maharashtra

In this section, we have tried to understand the economic context of seasonal migration. Attempt has been made to understand the economic circumstances of seasonal migrant households directly and indirectly. For the direct understanding of the economic status of seasonal migrant households, the researchers have collected monthly per-capita consumer expenditure (MPCE) and examined household's state of food security. Mean MPCE is estimated at Rs 1,483 (Median MPCE is Rs 1,414). From the distribution of MPCE, the lowest and highest per day PCE reported by the migrant households are Rs 19.4 and Rs 100, respectively. Of the total 340 households, 263 (77 per cent) households spent only a maximum of Rs 60 per person per day and the remaining 77 (23 per cent) households paid Rs 60 to a maximum of 100 per person. Additional to this, as background information, the researchers have collected information on land possessed by households. The research has found that 58 per cent of the households are landless. Most of the land possessed are barren (non-irrigated) and cultivable only during monsoon. For further explanations of causes of seasonal migration, the researchers have posed several questions to migrant households about their economic circumstances and employment opportunities in the village and nearby villages.

We collected information on perceived economic status of the migrant households in comparison with their non-migrant counterparts in the same village. When asked about their economic status with non-migrant households in the same village, 59 per cent of the households reported that there is 'no difference in economic situations'. However, 39 per cent of migrant households see them as 'relatively poor' when compared with the non-migrant households of their village. Six out of every ten households distinguish no difference in economic status between seasonal migrants and non-migrants in the same village. It is an indication of the fact that most of the non-migrants are not permanent non-migrants, they are mostly locked in the villages by circumstances. The majority of the households perceive that they are compelled to migrate due to their economic circumstances. Nine out of every ten seasonal migrant (88 per cent) households reported that they were compelled to choose the migratory option. To affirm this, researchers further asked that 'whether life is possible in villages if not migrating', 69 per cent of the households reported that 'life is not possible or somewhat not possible in villages if not migrated'. Three out of every four (76 per cent) households indicated that the primary cause of their migration is 'seasonal unemployment' in villages, and another 20 per cent said that 'less wage' in villages was the reason for their migration. 39 per cent of the households stayed back in villages (during dry seasons) at least once in the last five years. We have enquired about how they managed their subsistence when they were staying back? Of the households who have an experience of staying back at least once in the last five years were asked of their strategy of managing their subsistence. 45 per cent of them reported that they managed their subsistence by 'taking up works available in village/nearby villages'. Another 40 per cent of the households stated that they had 'borrowed money' to manage their daily living. Considering both the experiences, researchers have further tried to explore which coping method they feel relatively risk-free, and 95 per cent of the households said that taking up seasonal migration was a better option than staying back in villages.

Respondents were asked about their view on the economic status of the households who took up seasonal migration. They were given options, namely, 'the poorest', 'the poor', 'better off' and 'every households' and were asked for their opinion. In response to this, 96 per cent of the households reported that 'the poorest' and 'the poor' are the people who chose to migrate during dry seasons. Further, they were asked about their perception on the reasons for their seasonal migration. Response to this question affirms the fact that they were compelled to migrate due to their economic adversities. The most frequently stated three reasons for migration are 'to meet large emergency expenditures', 'indebtedness' and 'poverty'. The majority of the households reported that the large emergency expenditures (65 per cent), indebtedness (60 per cent) and poverty (59 per cent) were the primary reasons for their migration, whereas 17 per cent of the respondents reported that better wage at the destination was the reason for their migration. From this, it can be understood that by and large, seasonal migration is part and parcel of survival struggle of the rural poor. In conclusion, the reasons reported by the majority of the households are driven by a situation where they are unable to earn sufficiently in their villages to meet their basic needs. Thus, economic hardship drives people out of the village to sell their labour. It points towards the fragile nature of the livelihood available to the people in the dry areas of rural Maharashtra. Thus, seasonal migration from rural Maharashtra is distress driven in nature.

Box 12.1   Case study of a woman seasonal migrant from Beed district

Age 57 years

November 2013 at Ahmednagar

Subhadra Bhema Rao Gade is a 57-year-old seasonal migrant woman from a village of Patoada Tehsil of Beed district. She is an illiterate woman having eight children (seven daughters and one son). After her marriage, she started migrating for sugarcane cutting with her husband. Since marriage, she has been migrating to various districts of Maharashtra such as Pune, Satara and Ahmednagar. She is a Maang (Matang) which come from

Scheduled Caste (Dalits). Currently, she and her husband are living with their son's family which comprises of their daughter in law and three grandchildren. This time, the entire family (seven members) migrated to Ahmednagar district for sugarcane cutting. She said, 'This is like our second home, we have been migrating to this place from last 15 years and so.'

We have five acres of land in the village. The land is cultivable only during monsoon (Paanika bahut Samasya hai). We cultivate Bajra in the month of June and reap in the month of August after that cultivation is impossible. What we cultivate is not even sufficient for our use. We need to earn extra money through wage to manage our expenditure. We go for work to those who have irrigated land in our village and nearby villages. We work for them to plant Bajra, Toor (Daal), Kapus (Cotton) and Harbara (Chana). The wage for a male worker is Rs 200 per day and for the female worker is Rs 100 per day. Since there is an abundance of labourers and the land is non-irrigated, it is often difficult to get regular work for all. There are no alternative jobs available in the village or nearby villages. To sustain in the village, we compel to take an advance money from Mukhadam (labour contractor) of sugar factories. Mukhadam supplies labourers to sugar factories in various parts of Maharashtra and Karnataka. A koita (a group of two working people, usually a male and a female) gets an advance amount under the condition that they work for sugar factories under this Mukhadam during next dry season at least for six months. We have no livestock. In the village, we have a house and a land of five acres. As long as the document is with us nobody can steal it from us. We have nothing to lock and keep there. So we migrate when there is no job in village.

Subhadra's son and his wife (a koita) have received an advance of Rs 50,000. Subhadra's husband also helps them in their work so that they can repay the amount taken from Mukhadam as early as possible and free from the contract and earn some extra amount. Subhadra is not working for last few years due to her ill health. But, at the workplace, she has to prepare food, fetch water, collect firewood and look after her grandchildren. Her family is living near to a sugar factory where they have given a water tap for almost 100 migrant households staying there. But, she has to go to the hill (1 to 2 km) for collecting firewood. When asked about life at work place she said, 'As anybody else we are also comfortable in our village, but what to do, we have to live.'

When asked about the life in such a pathetic living condition she replied, 'this all comes after food, there is no meaning of having a good place to live without anything to eat and we have been living this life for years and habitual to these conditions'.

Source: Authors.

Box 12.2 A first-time seasonal migrant woman from Yavatmal district

Age 28 years

November 2013 at Ahmednagar, Maharashtra

Gaukarna Rajpatre, a Banjara woman from Yavatmal district, migrated the first time to work in a sugar factory with her husband. She is working for a sugar factory at Ahmednagar. She married eight years before, and she has no children. She is one among the 14 people migrated to this place from her village with a person of their same village. The agent is Mr Bosle. When asked about him, he said, 'I am a Marata and all these people are Banjaras, they all are from my village, and I took them to work. I paid each of them an advance of Rs 20,000 when they were in short of money due to no work in the village. They work for me without wage till their advance amount is repaid and after that I pay Rs 300 per day for two people.'

She is very thin and week. When asked about her pre-departure knowledge of work, wage and place, she said, 'I was not aware of the work, wage and place. I am coming first time from my house this much long.' To the question of the reason for migration she replied, 'We have no land, no water and no work there. When monsoon goes short, people from our place migrate. Otherwise, we work for those who have land.' She was not even aware of her next day's work. 'All the day we shift our workplace. Today it here, tomorrow it will be somewhere else, I don't know, wherever Mukhadam asks us to work we go there and work.' When asked about life at work place she said, 'we wake up 6 am and walk to work site start our work by 7 am. We have a break of total 1 hour for food on 10 am and 2 pm and stop working by 5 pm (9 hours of work). We go and prepare food and sleep by 10 pm.'

Source: Authors.

Consequences of seasonal migration

Though seasonal migration delivers a large share of subsistence to the rural household in Maharashtra, it bears severe consequences. The consequences of seasonal migration are presented in terms of the people left behind, child migrants and child labour, working and living condition of migrants and fear and insecurity experienced by seasonal migrant women. Each of these aspects is discussed in this section.

People left behind

More than half (55 per cent) of the total population who remained at home were children and 10 per cent were elderly. Some of the families took their children along with them to the place of migration and some others left them in villages. They mostly kept the school-going children in villages so that they would not miss their school days. Children were staying alone when their parents migrate. 24 per cent of the total studied households have reported that they kept their children alone in their villages during the last migration.

Child migrants and child labour

As far as seasonal migration is concerned, children are an important component of the work unit (Koita). Seasonal migration takes place under the Koita system wherein a work unit comprising a husband, wife and one child or two children. These children usually help their parents in their work and look after their younger siblings. When asked about her work (a girl aged 14 years working in a sugar factory of Ahmednagar district with her parents), the girl replied, 'my sister and I help my mother in her work, because, the volume of work matters. If you work more, you get more'. While interviewing a school teacher on the reason why many children in the villages were out of school, she replied, 'many migrants take their young children to the workplace so that they can look after their siblings while the parents are busy working. These children miss many of their school days and eventually many leave school in the middle of a course of study to become labourers'. Many children begin doing small chores at the work sites and eventually become absorbed into the general labour force. The Koita system places the children in hard labour that jeopardises their lives, safety and physical or psychological development.

Among 604 children in the community which is 32 per cent of the total population, 47 per cent (282) of children age 0–5 years and the remaining 53 per cent (322) of children are at school-going age. In children of school-going age, 14 per cent of the eligible children have never been to schools or dropped out of schools. The underperformance of students is estimated by matching their age with standard (class) in which they study. If the age-standard combination matches with which standard they are supposed to study, those students will be considered as performing students. If the standard in which they currently study lags behind the standard in which they are expected to study as per their age, those students will be considered as underperforming students. For testing this, the researchers have considered all the children (6–14 years of age) who were going to school at the time of survey. There were 277 children in this category and 163 children were found studying in a standard they were supposed to study and 113 children (6–14 years) were studying in a standard which lagged behind the standard in which they were supposed to study as per their age. Thus, it can be drawn from this analysis that 41 per cent of these children are underperforming in their studies. The researchers have considered only those people in the age group of 16–19 years and checked their academic achievements. This has been taken up to understand the educational achievement of the people in the study population. There were 95 children in this age group. It has been found that 19 per cent of them dropped out of school after a maximum of seven years of schooling. Another 63 per cent of these children were studying in eighth to tenth standards. Only 18 per cent of this age group successfully completed tenth standard. It is also found that 31 per cent of the children of school-going age accompany their parents while they migrate and in which 63 per cent of children work (help their parents in the workplace). About 41 per cent of the total migrated children were of school-going age.

Box 12.3   Case study

School drop-out children of seasonal migrant woman

Sheela Laxman Adagade, a seasonal migrant woman to sugar factory of Ahmednagar has three children, Sheetal (14 years), Rekha (12 years) and Sandeep (9 years). All the three children stopped going to school due to seasonal migration of their parents. Sheetal a seventh standard passed, dropped out of school from eighth standard after losing her interest in studies. She has been taken by her parents whenever they migrate because nobody was there in the village to take care of her and her siblings. Her temporary absence (*five to six* months in an academic year of ten months) made her an underperforming student in her class. She lost interest in studies due to her poor performance and blaming attitude of teachers. The younger brother of Sheetal said, 'they don't like us, they punish us in class'.

From last three years, Sheetal's family has not gone back to their village. 'We have only a house there that too made with tin sheets. We don't go back to the village. My father has so many friends there. Once he goes there, he starts drinking and makes problems to all. We are happy here and live in a rented room for six months which is far better than the house we have in the village. Here my father goes for construction job and mother works as agricultural labour. My sister and I help my mother in her work, because, the volume of work matters. If you work more, you get more. When sugar factory starts functioning (sugar canes are ready to cut), we shift to this place for six months. We all work here. My sister and I manage household activities, and father and mother work. I don't have any interest in continuing studies. My sister Rekha stopped going school when she was in 5th standard. Many a time we had to live in starvation.'

Source: Authors.

Living and working condition and wage of migrants

Seasonal migrants are forced to live in vulnerable circumstances due to many reasons. Since they cannot afford a rented house, and to save transportation cost and time which they can utilise for work, they prefer to stay near to their workplace. They are deprived of many of the basic amenities such as a proper place in which they feel safe to sleep, toilet and access to safe drinking water. The vast majority of the migrant households (97 per cent) live in temporary huts near their workplace, 98 per cent report that there is no provision for toilets and 49 per cent report that they do not have access to safe drinking water.

Workers are paid a part of their wages during the employment period on a weekly basis, and the remaining amount will be settled at the end of the season. This binds the worker to the work during the period, whereas the carry-over of advances from one season to the next ensures their availability for the next season. The work usually entails long working hours. Seasonal migrant workers are found working an average of 12 hours a day and are paid an average wage of Rs 219 per day. About 56 per cent of women report that they work 12.6 hours a day, and 37 per cent of respondents report that they work more than 12 hours a day. Eventually, 93 per cent of the total workers work 12–16 hours a day at the workplace. When asked about the number of working days in a week, 76 per cent of workers report that they work all the seven days in a week.

Fear and insecurity experienced by women seasonal migrants

In this section, we deal with the experience of seasonal migrant women regarding safety, fear and violence. For which several questions were posed to them such as life conditions at the workplace with respect to village (how do you rate the life conditions at the place of work in comparison with your village?), safety at the destination (where do you feel more safe to live?), response of labour contractor to make the place safer (did the labour contractor/factory take any measure to make the living place safer?) and experience of violence (have you been subject to any type of violence?).

Of total women respondents, 57 per cent feel that there is no difference in their living conditions between the workplace and the village. However, 23 per cent of women feel that the living conditions available to them in the place of work is worse than that available to them in the village. Remaining 20 per cent of women feel that workplace is better than that of their village. As far as safety of seasonal migrant women in working and living places is concerned, 87 per cent of women respondents report that they feel safer in their villages than in the workplace and 45 per cent report that they are really afraid of living in the workplace, 98 per cent of women respondents reveal that no safety measures have been taken by either labour contractor, employer or anybody else.

Even though it is not easy to capture violence against women and the depth of their experience in quantitative terms, we have asked them whether they were the victims of verbal, physical and sexual abuses in the workplace during their last migration. Only 3 per cent of migrant women reported that they faced verbal abuse during their last migration. The proportion of women who report physical violence and sexual violence during their last migration was 1 per cent and 0.6 per cent, respectively. In conclusion, it is found that women migrant workers are forced to live in vulnerable circumstances, and there is no mechanism in place to monitor the violation of human rights and labour laws. Their necessities are not met. In the case of seasonal migrants, they work hard and lead a very low-quality life at the destination. Their working place seems like 'no law jurisdiction' since there is no application of labour laws or no system to monitor the violations.

Discussion and conclusion

This paper is based on 340 seasonal migrant households with a population of 1,881 persons from 14 randomly selected villages in Beed and Solapur districts of Maharashtra. 75 per cent of these households have a history of migration during all the dry seasons of last five years and 83 per cent have committed to migrate by receiving an advance amount from labour contractors during the dry season of 2013. Seasonal migration from rural Maharashtra is a process in which people periodically leave their permanent residence in search of wage employment, stay away from the home for a period of up to six months and then return to their homes. It is a massive event that occurs in every dry season (November to May), where households move to better-endowed areas.

The people who take up seasonal migration are mostly SC and ST households. It is also found that 69 per cent of the migrant households are either landless or having only marginal land of less than one acre. These prima facie facts lead us to assume that seasonal migration from rural Maharashtra is mostly by the people of lower social and economic strata. Seasonal migration is short term and repetitive in nature and adjusted to the annual agricultural cycle. People leave their villages between September and November every year and come back to the villages between February and April. Seasonal migration from rural Maharashtra is temporary in nature and over half of these migrants move to Karnataka and others circulate within the boundaries of Maharashtra. These circulating labourers are mainly absorbed by sugar factories for sugarcane cutting and transportation and by brick kiln for brick manufacturing.

In 2012, 65 per cent of people from the study households migrated seasonally across the district or state lines for employment. Of every 100 seasonal migrants, 20 were children (0–14 years), 75 were working-age population (15–59 years) and 5 were elderly (60+ years). Households near absolute dependency on agriculture, and almost non-existence of other significant sources of employment make life difficult for people in the study areas. To reduce the risk emanated from income fluctuations mainly caused by seasonal unemployment, primary income earners of households temporarily migrate to outside of their usual working areas to smooth out income flow. The majority of households (76 per cent) reported that 'seasonal unemployment' in villages pushes them to take up seasonal migration. These households find seasonal migration as a relatively risk-free coping strategy against the livelihood crisis which occurs in every dry season even though it has heavy social cost. Thus, the households in the dry areas of Maharashtra take up seasonal migration to manage the income shortfall due to the inability of their villages to provide them with a reasonable subsistence.

Overall, it appears that seasonal migration takes place mainly for survival and repayment of debts and that a large proportion of their earnings from migration is utilised for day-to-day expenses. Thus, rather than a demonstration of an attractive alternative livelihood, seasonal migration is a last resort for the poor people of rural Maharashtra to earn a livelihood. This study has found that seasonal migration is a critical component of the livelihood portfolio of the studied households. It is found that a male worker derives 77 per cent of his annual wage income from seasonal migration, and it is 83 per cent in the case of a female worker. At one hand it indicates how critical is seasonal migration for the survival of these households, and, on the contrary, it shows the bleak prospects and opportunities for the livelihood at their usual place of residence.

Though seasonal migration delivers a large share of subsistence to the studied households in rural Maharashtra, it bears severe social costs. The cost and risk of seasonal migration are found heavy, including the risk of not being able to send children to schools. More than half (55 per cent) of the left behind population in the villages were children, and 7 per cent were elderly. A quarter of the total surveyed households left only children in villages during the migration season in 2012. This points to the fact that there is a tendency to leave behind economically unproductive population in villages. It is found that seasonal migrant workers were forced to live in vulnerable circumstances, and it puts a greater burden on women seasonal migrants. Many eligible children in the study population are found out of school, and many migrate and thereby lose school days and put to work at the destination. Nevertheless, these people are willing to incur these costs or risks to ensure their survival.

In the destinations, seasonal migrant workers work an average of 12.6 hours a day. Most migrant workers work in the scorching heat, six days a week. Children who migrate with their parents help them in their work and ultimately join the workforce. As a concluding remark, the nature of annual seasonal migration is predominantly debt-bonded, harsh working conditions, and disrupts children's education. Even though these households' survival is ensured by seasonal migration, possibilities of social advance for them are very limited due to the heavy social cost associated with.

Based on the findings, we suggest policy measures commensurate with the ecological background of these districts intending to protect the livelihood of the poor. Economic and social backwardness is inherent in the study population. Serious efforts need to be taken by the governments to incorporate the poor people in the process of development. The regional imbalance and poor incorporation of the socially backward population need to be addressed immediately. Mere implementation of the schemes to address the repercussions of the real problems that are largely concealed will not solve the problem. It is not seasonal migration that needs to be checked, but the underlying problem where the root lays needs to resolved. The entire dry region is not suitable for water-intensive cash crop like sugarcane. Due to wrong policy measures of expanding cash crop cultivation and industrialisation, this region is facing acute shortage of water, leading to further deprivation of the marginal cultivators. Rather, policy must focus on basic food crops that are congenial with dry farming. It can help people to feed their family round the year instead of earning a mouthful for a short span of time and then migrating to a destination for sheer survival. In such migration, neither have they been able to gain economically much nor does it enhance the social cost. Creation of small water reservoirs, watershed management, efficient water distribution policy, rainwater harvesting, dissemination of knowledge on dry farming, horticulture, incentives for livestock rearing and huge public-private investment are required to address poverty and inequality for the development of dry rain shadow areas of Maharashtra.

References

Breman, J. (1985). *Of Peasants, Migrants and Paupers: Rural Labour Circulation in Capitalist Production in West India*. Delhi: Oxford University Press.

Breman, J. (1996). *Footloose Labour: Working in India's Informal Economy*. Cambridge: Cambridge University Press.

de Haan, A. (2002). Migration and Livelihoods in Historical Perspective: A Case Study of Bihar, India. *The Journal of Development Studies*, 38(5), 115–142.

Deshingkar, P. (2010). *Migration, Remote Rural Areas and Chronic Poverty in India*. Working paper 323. London: Overseas Development Institute.

Deshingkar, P., & Start, D. (2003). *Seasonal Migration for Livelihood in India: Coping, Accumulation and Exclusion*. Working paper 220. London: Overseas Development Institute.

Gidwani, V., & Sivaramakrishnan, K. (2003). Circular Migration and the Space of Cultural Assertion. *Annals of the Association of American Geographers*, 93(1), 186–213.

Haberfeld, Y., Menaria, R. K., Sahoo, B. B., & Vyas, R. N. (1999). Seasonal Migration of Rural Labour in India. *Population Research and Policy Review*, 18(5), 471–487.

Narayanamoorthy, A., & Venkatachalam, L. (2011). Farmers' Right to Water. *The Hindu, Business Line*. 7 September 2011.

Planning Department (1984). *Report of the Fact Finding Committee on Regional Imbalance in Maharashtra*. Government of Maharashtra.

Rao, U. (1994). *Palamoor Labour: A Study of Migrant Labour in Mahabubnagar District*. Hyderabad: C. D. Deshmukh Impact Centre, Council for Social Development.

Rogaly, B., Biswas, J. Coppard, D. Rafique, A. Rana K., & Sengupta, A. (2001). Seasonal Migration, Social Change and Migrants' Rights: Lesson from West Bengal. *Economic & Political Weekly*, 36, 4547–4559.

UNDP (2012). *Maharashtra Human Development Report*. Pune: Yashwantrao Chavan Academy of Development Administration & New Delhi: Sage.

13 Migrant ecology

Shareena Banu C. P.

People migrate from rural areas to urban centres due to various reasons. One major reason is development-induced destruction which leads to losing not only land but their original ecology also. The main cause of today's forced migration is the whole ecological balance getting disturbed. In contemporary times, it is the most pressing issue and this process is described here as a new concept called 'migrant ecology'. It is conceivably both the original habitat in which rural people depend upon their livelihood and the new ecology into which they are forced to shift which is the urban life. When their rural ecology is destroyed due to developmental reasons, they migrate in search of new avenues. In the urban system they are the most marginalised and live a life which is completely alien to them.

Presently, India is facing forced migration as a major problem. But we pay scant attention to what actually leads to migration. It is not just their livelihood but the whole rhythm of life itself that is disturbed. This perspective is generally missed out. The rural population no more has the faith that their livelihood would be protected. Out of helplessness they migrate, leaving their own habitat. Displacement has become an everyday reality of rural population. Hence we propose 'migrant ecology' as a perspective to deal with this situation. Migrant ecology is an idea for a better ecological perspective that protects the livelihood of rural population. It looks at forced migration as something which needs to be addressed. It is necessary to regain the lost confidence of rural population and protect their ecology. Ecology here refers to not only equilibrium in nature but also a state of mind. It is a fluid concept. Rather than confining it to geographical location we consider it as a spatial category. The sense of loss is intensified in urban space without providing a better and balanced life.

There are three sub-themes that form a part of the construction of the new concept of migrant ecology. They are (1) from a social ecology perspective theme to migrant ecology, (2) ecological movements in India and (3) marginalised by development and the question of livelihood.

Part I

From a social ecology perspective theme to migrant ecology

Social hierarchy and ecological crisis

Man appeared on earth much later after other life forms had made their signature. Among other common descriptions, the expression 'mother earth' implies the ultimate life-giving force and the symbol of tolerance and endurance. Its capacity to produce and reproduce life forms and its power to self-organise itself has always remained an enigma for theologians, geologists and natural scientists. Primitive man worshipped nature and elevated its meaning beyond mere material existence. Nature was seen as made up of elements such as earth, air, fire, ether and water. Man was also believed to be created out of these five elements.

Over a period of time, with the spread of industrialisation and modernisation, man's conception of nature changed. It started with exploration of the world through expeditions and gradual exploitation of raw materials and riches. As a result, the eastern parts of the world were colonised by western industrialised countries primarily for the extraction of natural resources. The more the technology advanced the greater the exploitation of nature. In the post-industrial world, ecological imbalance has by now reached an alarming stage that our future generations would not have any chance to enjoy the riches of nature as our predecessors did.

Drawing upon Murray Bookchin's (1921–2006) idea of social ecology we develop a new argument on migrant ecology. He was one of the important thinkers of our times who devoted himself to study human's relationship with ecology. He was one of the pioneers of the worldwide ecology movement. The basic premise of his study was the problematic relationship between nature and society. It was postulated that most of the ecological problems we face today are the result of persistent social problems existing among human beings.[1] He points out, 'The way human beings deal with each other as social beings is crucial to addressing the ecological crisis. Unless we clearly recognize this, we will surely fail to see that the hierarchical mentality and class relationships that so thoroughly permeate society giving rise to the very idea of dominating the natural world.'[2]

From one level, modern technology is a manifestation of man's ways of using science but on another level it is a source of our understanding of man's relation with other men. We often fail to address the real dangers of our society in which man's desire to overpower others by maximizing one's capacity for control and venturing into powerful commercial and profit-making development projects which inevitably lead to mass displacement and migration. Narmada Bachao Andolan (NBA) is a glaring example of this kind. Scholars like Bruno Latour have identified some basic problems of modernity in this dualistic distinction between nature and society.[3] Nature and society co-produce each other in its enduring relationship. Any form of reallocation and independence, as is visible for instance in the separation of social and natural sciences, would ultimately distance us from our own ecology.

Neo-Marxist thinkers like Bookchin also raise the question of the irrationality in our attempts to address the ecological crisis without accepting and addressing the need for change in the existing value system. There cannot be two movements: one which addresses environmental issues and another to address a different set of social issues. Dissociating between the two movements means that there has not been much change in our perception of our environment. The 'social ecology' argument, therefore, envisions an all-inclusive change in our society. This is only possible if we start addressing the ill effects of a market society which has its basis in domination of man by other man. The established mode of relationship stakes the humane attitude to the world and such a mentality would only fulfil the power and wealth motives of the rich. The lack of symmetrical co-existence has had far-fetched effect not only on our social life but on our ecology also.[4] On 5 December 2015 we observed World Soil Day yet 'forced migrants' problem of livelihood and the need to preserve their ecology have never come to the limelight.

Civilisation and man's estrangement with nature

Only by way of re-conceptualising the existing social relations could one re-imagine the establishment of a strong and vibrant ecological society. Since we have stepped out of nature in the process of fulfilling ourselves, the re-entry into the natural world is a prerequisite.[5] The human world has to reunite and merge with the natural process of evolution and preserve its diverse vegetation and species. It is true to observe that the history of civilisation is perhaps the history of man's estrangement from nature. The retreat to nature is a possibility only if the decision is made by conscious human beings. For instance, it could be similar to the organic society of the past which exhibited strong communal ties binding people together. An ecological society is the one which preserves the notion of universal humanity of our culture. It would be a free society wherein social hierarchy and domination would be replaced by a deep sense of interdependence and mutual co-existence, so much so that it would fulfil the biological need for care, cooperation, security and love. These are the essential qualities of human nature and would have been ideally imbibed through socialisation and group life. These qualities are also essential to the evolution of human subjectivity and personality. Preliterate societies and modern states represent different ecological sensibilities. The latter represents property rights, considering nature as a resource, largely propelled by acquisitive impulses, usufruct and a system of distribution that hardly knows the virtue of exchange. Humanity has to become that medium which echoes nature's voice, survival quest and its fecundity. This is the desire of most agricultural migrants of India. They desire to go back to once own ecology, roots and culture.

Modern civilisation presupposes that the autonomous individual is the one who is 'competent and therefore capable of making rational judgments; in short, the individual is capable of functioning as a self-determined, self-active, and self-governing being'.[6] No migrant man can make a choice of staying back in his land when traditional employment is at stake. His idea is to protect the well-being of his family and preserve the basic means to find a day's meal. Therefore the rational migrant man cannot anticipate a comeback to his original land. The sense of collectivity parallels an intuitive ecological sense of wholeness, stability and fecundity. Building up an ethical character therefore is the most important social role. Afforestation drive, recycling of organic waste and industrial ecology could be a starting point in this direction of returning to nature what primarily belongs to nature. Human beings are neither the custodian nor the patriarch of nature. By way of adopting the most eco-friendly method of generating energy, man could save the ecology for the children of our present life time.[7]

Social hierarchy in society is a reflection of the presence of a system that neglects the idea of diversity and human differences. It is ultimately the human virtues that are reflected in the idea of migrant ecology. A society which cannot protect humanity will not be in a position to give itself to the cause of ecology. It is evident from history that the two World Wars have caused not only widespread calamity but massive destruction of the ecology also. More and above all, the violent exploitation of nature for development causes severe damage to the migrant ecology.

Emerging new ecological consciousness

Man is a species being and he cannot be dissociated himself from other species. Any attempt to cause damage to humans cannot be committed without ecological damage. The carpet shelling and bombing targeting human occupation is another feature of modern warfare. Such violence not only leads to the annihilation of human life but also extends to the loss of marine life, forest, vegetation and more, and above all the destruction of the ecological balance of a region. Both the living and the non-living organisms are therefore at the mercy of modern human civilisation. It takes long years of painstaking labour to revive the ecological balance of a region. Nature's way of healing its wounds might take more than millions of years. This is more so in the case of people who are evacuated at mass level and rehabilitated at a different place.

It could be purported that nature is the best judge when it comes to natural diversity and it fulfils the meaning of the fecundity of life by the celebration of the diversity of different species. Nature acknowledges the singularity of each being but at the same time each and every organism is interconnected at a deeper level to the larger ecosystem. The natural evolution of species, which is an ongoing process, leaves the message and the reminder to humans that as a species it should acknowledge racial and cultural diversity and learn to accept each other. Every migrant teaches this. The knowledge that could be imparted that treats nature as the basic pedagogic ideal is yet to be explored. It is also true that the cultural and racial difference among human beings is also a result of the various modes of relations and engagement with nature. Nature has played an important role in predetermining and coding the racial features, and also marking the ecological origin of food habits and social relationships between human beings. Every primitive agriculturist is aware of it and when they become migrants they face problems to adjust to urban settings.

Migrant ecology is an ideal for a futuristic free society that conceives of a reconciliation and a return back to nature. We have no choice but to redefine the prospect of our civilisation in tune with our ecosystem. The heralding of a new ecological society would be making a re-harmonisation with nature. This re-harmonisation with nature would mean to strike a liveable code between human and human; whether rural or urban. This would in turn bring a new ecological sensibility that has achieved a desired distance with most social evils and a range of activity which degrade man. It is necessary to be humble enough to surrender oneself and fall back into the lap of nature. The dialectical engagement of man with the external world was certainly predetermined by the very process of human coding which makes us nothing but another organism with certain added mental attributes. A rural migrant's sense of estrangement from nature is a permanent one, since he comes to urban space with his state of mind as rural. He has many family members and relatives still living in the rural areas and has the cultural heritage exhibited like a kind of extreme and intermediate rurality.

Arguing on the lines of Bookchin, rendering a new method of analysis of the ecological crisis ought to be an ontological one. This objectification of the spontaneity induces a discord with oneself and one's social and natural surroundings. Migrant ecology therefore envisions more than a mere method but something that addresses the question of how to deal with the destructive ecological reality of our times.

The endemic anxiety and dissociation of man should be replaced by a new ontology of substance based on the self-organising principle of nature. Hence, the new ecological consciousness would require the reconsideration of existing contours of our thoughts on the relationship between nature and culture. It is in this context that we have to discuss the life of our ancestors who have traversed much and could follow the path getting deeper into the recess of nature. Science has not been able to replicate but burden man of his own knowledge on matter, mind and nature. Scientific endeavours have overlooked the limits and impact of human exploitation of nature and its diverse resources. Therefore we need to uphold a process approach towards ecology to ensure the safe retreat of man back to nature. Current dialectical framework forged between thought and matter, mind and body, subject and the object and nature and culture needs to be reargued by bringing back the safety of migrant ecology as the first priority. Science has to be therefore intrinsically ecological and life nourishing, and a viable source of exploring and preserving the vital character of life. The basic premise of science should therefore be life itself.[8]

Conceptual preference for ecology over environment

The term ecology and environment are used interchangeably in public discourses. But there has to be a semantic distinction drawn between the two.[9] Environmentalism designates a mechanical outlook towards nature. It has a very static view and endorses a classical materialist conception where man and nature do not share a common ground but subscribe to a view of holding distinct history of its own. This is conducive for a capitalist economic system which subscribes to a minimalist view of migrant ecology. By way of transgressing this balance, the profit-oriented dominant interest groups mistook the need of the consumerist world as the need of the species. As a species being, man seeks vibrancy and creativity and not just the fulfilment of basic materialistic needs. The notion of the creation of a green habitat around us could be seen as a viable proposition but it is nothing short of mere tokenism, as it considers natural settings as passive pieces of art form. It does not either render itself amenable to address the question of human greed and insatiability as nature is seen to be of ornamental value and to be used for its use value especially for raw materials.

Change does not come from the above but by the transformation of subjectivity and sensibility. Environmentalism does not scuttle the existing development paradigms on migrants, nor does it intend to provide an alternative worldview. The real crisis and the magnitude of the issues are underplayed to such an extent that the capitalist logic of mindless production and consumption remain unattended. After arriving at its dismal low, it would not suffice just to say that by way of facilitating the inroad to eco-friendly technology and production process one could bring substantial changes in the human desire for luxury and comfort. The basic premise of the debate still continues to be the exploration of the ways and means to dominate and exploit nature or migrant ecology.

Therefore 'ecology' must be distinguished from 'environment'. The term environment is vague and abstract as it would mean only the immediate presence of natural setting accessible to the urban population. Ecology instead does not just indicate the arbitrary human accommodation of the natural world. The usage reminds us of the relevance of nature and to integrate it to our very definition of society itself. For instance, the equilibrium of nature enables humans to settle even in the most difficult Himalayan territory. The life of human beings in such mountain terrain means to live in the hinges of existence by way of accepting the interdependence of humans on nature.[10]

Many scholars now show preference for the term ecology instead of the earlier common popular usage of the term environment. The term environment has a limited purview. Instead ecology brings into perspective the unequivocal interrelationship between different species and natural things. The things of the natural world such as birds, ants, mice, domestic and wild animals and organic and inorganic objects migrate and are directly intertwined with each other.

The term environment hesitates to declare the intercourse between migrant ecology and society; rather it sustains a mode of exchange that is instrumental and dissociative. It rather rationalises the truce between the two entities. The lasting equilibrium cannot be sustained until the order of things in the natural world is not brought back to its original condition of existence. Environmentalism approaches these issues as the distilling of the impurity generated by modern technology and industry. It believes only in making gardens, and collecting and removing waste materials and toxins from human habitat. It intends to employ new environmental-friendly machines and tools which mute the ordeal of the current upset. Environmentalism does not question the most basic premise of the present society, notably, that humanity must dominate nature; rather, it seeks to facilitate that notion by developing techniques for diminishing the hazards caused by the reckless despoliation of the environment.

Hence migrant ecology approach as a whole has a dynamic based on the concept of unity of diversity. The capacity of an ecosystem to retain its integrity depends not only on the uniformity of the environment but also on its diversity. It is this ever-changing difference between ecological spaces that we see the earth coordinate the functioning of its cycles and rhythm. The rhythm of nature is co-existence, not retribution. Many of the modern diseases are a product of the food we eat. Artificial farming and use of chemicals have also reduced the fertility of the cultivable land, which led to the widespread migration of peasants to urban areas. By bringing back the ecological stability through simplicity and thoughtful intervention we could regain physical and mental harmony. One needs to conceive an ideal before us in order to surpass the existing hurdles coming in the way towards an ecological society. A migrant ecological community can achieve a balance of co-existence with nature and mean a collective reunion of man back into the cycles of nature. Moving in this direction of a migrant ecology would supply the necessary reserve to bridge the vital gap existing between humans and nature.

Part II

Ecological movements in India

Contextualising the causes of ecological movements in India

There is a deep concern emerging about the changing condition of ecology all around the world. India is one of the worst affected regions in this regard. We can understand the intensity of the ecological issues from the perspective of various ecological movements in the country. One of the social aspects of the ecological issues in India is the question concerning the availability and equal access of necessary natural resources such as water, land, fodder etc. We could say that natural resources of a state or a country as a whole should be utilised for the benefit of everyone but on one level there is an uneven pattern of distribution of natural resources developing among its real beneficiaries. And, on another level the state in its claim for development is taking away the ownership of forest and agricultural land. The forest and the cultivable land are not mere natural resources but also a question of livelihood and a habitat for many. Hence, this has badly affected the lives of millions of tribals, peasants and particularly women and small farmers. They are the major population constituting the current migrants of India.

India has a forest policy which is blind to the rights of the forest dwellers who are the real inhabitants of these natural abode. The use of forest resources by the tribal people has been a natural right for which they had prehistoric claim which is now under the control of the state. For example the word 'Jharkhand' means the 'land of forest'. The state's adivasi heroes such as Tilka Manjhi, Sidhu, Kanhu and Birsa Munda struggle for their land. However, now Jharkhand is one of the major migrant source states in India. The real issue is therefore the irresponsible and continuing interference in the life of the poor and the disruption of their peaceful and harmonious intercourse with nature. It raises deeper issues of governance and massive violation of ecological guidelines regarding rehabilitation, appropriate compensation and the urgent need to evolve socially sensible and sustainable development projects.

In the context of India, the state has always been the key agent of the development process. It is the responsibility of the state to win the confidence of the people in new development projects. However, the increasing amount of dislocation and land alienation in the name of development initiatives has raised serious doubts about the state's role in the protection of ecology and taking the responsibility of the social and cultural impact of mass displacement migration. While the state through its policies urges for sustainable development, it has also given open invitation to market forces to blindly extract minerals and natural resources of the country. This is particularly true when one evaluates the lack of commitment and the arbitrary nature of government agencies in implementing environmental laws.

The development model in India was mostly based on the modern western mode of industrialisation. Most governments had no vision or plan with regard to ecological preservation. This is the main reason why the capitalist market entered the field of resource extraction with such an ease and thus made it unanswerable to the cause of environment. The conflicting relationship of people with development agencies took shape in the form of various people's movements which mostly included the marginalised sections of the society. This is particularly so in the case of migrants because they are the marginalised sections who were the real victims of deforestation, displacement and land evictions. Therefore, the common masses are left with no option but to mobilise and encounter the market forces directly without any state support.

There are issues which are purely ecological: some are development-induced ecological issues and some others are purely social issues of development which are directly connected to the environment. While evaluating the ecological movements in India we could see that they have questioned both the direct attack on agricultural land and peaceful settlements and the extraction of natural resources. Large-scale deforestation and air and water pollution due to construction of dams, factories etc. are the major points of contention. Indirectly one could argue these are the main causes of rural-to-urban migration.

Some of the noted and well-known ecological movements in India are as follows: Chipko Andolan, Garhwal; Save the Bhagirathi and Stop Tehri project committee in Uttar Pradesh; Save the Narmada Movement (NBA) in Madhya Pradesh and Gujarat; Youth organisations and tribal people in the Gandhamardan Hills in Orissa whose survival is directly threatened by bauxite deposits; the opposition to the Baliapal and Bhogarai test range in Orissa; the Appiko Movement in the Western Ghats; groups opposing the Kaiga nuclear power plant in Karnataka; the campaign against the Silent Valley project in Kerala; The Rural Women's Advancement Society (Gramin Mahila Shramik Unnayam Samiti) formed to reclaim waste land in Bankura district in the State of West Bengal; the opposition to the Gumti Dam in Tripura. Here we evaluate the nationwide causes, effects and success of two ecological movements in post-independent India: the Chipko Movement and the Narmada Bachao Andolan. Except a few, most of these movements have been witnessing the victims being forced to migrate from their land.

Chipko Movement, Garhwal, 1973

Early 1970s was still the period of Gandhian values which continued to hold a sway over the people as a method of political resistance. With its sheer simplicity and deep effect it continues to bring people together and endure the capacity to transform the people involved in both sides of the struggle. The people of Garhwal Himalayas in the State of Uttarakhand (then in the state of Uttar Pradesh) when faced with the issue of rampant deforestation in the area adopted the Gandhian method of Satyagraha and non-violence. To counter the felling of trees by the private outside contractors, who used government licence for commercial extraction of the forest, the people in the area came together in small groups and created small human chains by hugging (Chipko) the trees, when the timber merchants arrived on the scene. The movement was led by Gandhian leaders such as Chandni Prasad Bhatt and Sunderlal Bahuguna and peasant women, who had gradually started working on the rights of the local small-scale extractors, further extended their struggle to the larger labour and development issues of the Himalayan region.

The decline in the employment opportunity has led to the mass migration of the hill population to the plains in the 1960s which was a major event associated with the Chipko Movement. The local youth migrated in large numbers. The vision of the locals gradually changed and many started reconsidering subsistence on their local ecology. Local leadership checks the effect of this outmigration.

It is a landmark event in the history of ecological movements in India in more than one way. It shows the beginning of the modern era of environment awareness in the context of India which had its indigenous roots and local genesis. It was a movement which was the first of its own kind in India and is also marked by an active participation of the women folk of the region. The peasant women of Reni village and their fearless encounter with the state authorities, especially with the forest department, stood as a symbol of how culture and traditional wisdom shared by the people manifested in the form of divine and mythical connection to mother earth. That became their source of ideology to protect their environment and their source of knowledge of the local ecology.[11] Not only in this case, but in the subsequent years also, when the movement spread to other regions, it was the peasant women who came to the forefront to reclaim their natural right to forest especially fodder and water. Women from other districts in the region carried forward the struggle in similar manner of protest when the loggers came to axe the timber trees. It soon received nationwide attention and the people across the country at grassroots level started raising their voice against the deforestation under state protection. This happened especially when their male counterparts made up their mind to move towards the cities for better job prospects.

Taking inspiration from the movement, it has spread out to other forest areas of Himachal Pradesh, Rajasthan, Bihar and Karnataka.

Women's participation upholds the precedence of non-violent protest against ecological degradation in India as a whole. The whole theorisation of the relationship between women and ecology has developed into a separate field of study called ecofeminism. It is pertinent here to note that J. Bandopadhyay and Vandana Shiva, prominent ecological theorists of our times, call the Chipko Movement as a civilisational response to the forest crisis in India. It has become an exemplary case of grassroots environmentalism in the Third world.

The United Environment Programme Report observes that the Chipko Movement was a revolution of its own kind to preserve the forest resources. It still inspires the new generation ecological activists who are persuaded by the power of the people's movement and the capacity of the masses to critique the irrationality of the development policies which is blind to the question of ecological conservation. The Chipko Movement has stood for the cause of the protection of forest land in the Himalayan region which is a very sensitive geographic terrain whose green cover is depleting, causing soil erosion and water shortage. The mountain regions, unlike the plains, are mostly neglected because of the lack of transportation and other infrastructure facilities such as electricity and water. Because of the difficulty of the terrain it has hardly invited the interest of the subsequent governments to take genuine interest in addressing its economic problems.

During this period the region witnessed mass migration. 'Not only do females in the Chamoli district out number males by four percentage points, but also the single-member female households out number single-member male households. The majority in these single-member households belong to the 50-plus age group. Male migration from the hill areas to find work in the armed services and other jobs in the plains is fairly common, with women left to look after land, livestock and families.'[12]

The Chipko Movement has a major influence on the forest policy of India, especially the passing of the Indian Forest Conservation Act of 1980.[13] There was widespread concern of the depleting forest resources of the country. '. . . ecological scholars and activists argued that nearly 4.3 million hectares of forest area in different parts of the country had been deforested by state governments within 25 years (1950–1975) under the pretext of promoting industrial development and hydro-electric projects.'[14] It has reminded the policymakers that though people suffer from poverty, their consideration for nature and environment cannot be negotiated in the name of commodity production. The deep-seated philosophy of life of an average Indian is deeply grounded in an eco-friendly imagination of the harmonious interface between nature and culture. The Himalayan Mountains has remained a sacred geography in the minds of the average Indian. The development agencies have underestimated people's sense of connectedness to their natural environment; they could view it only in terms of economic investment and the capitalist logic of blind resource extraction and exploitation. It was a movement which brought to light the deep and matured ecological consciousness of the Indian rural population of which Gandhi had so much confidence and faith in as the cradle of India's wisdom and civilisational glory. It exposed the fundamental tension existing between modernity and culture, and between development and migration.

Narmada Movement against dam construction in Madhya Pradesh and Gujarat

Construction of large dams in India as part of development initiatives has an alternative history. It is the history of modern displacement of people from their original ancestral land due to developmental planning in India. Narmada Movement is connected to the plight of people affected by the river valley projects especially on the banks of river Narmada in western India. The Sardar Sarovar Dam Project, the second largest in the world, was to affect two million people spread across a total of 248 villages near the valley cutting across three states in western India – Madhya Pradesh, Maharashtra and Gujarat. In number, the people who are either affected or displaced by dam construction in India were estimated to be more than 20 million. The most affected category of population is the tribals. The extent and magnitude of the effect of displacement is beyond measures.[15] Medha Patkar, one of the prominent leaders of Narmada Movement initiated the anti-development protest of the tribals. Among other things, one of the main demands of the NBA is the rehabilitation from the perspective of a whole community rather than providing individual compensations.

   The issue of displacement is debated from different corners. One of the early initiatives in this regard was carried forward by the Chhatra Yuva Sangharsh Vahini (Student-youth struggle force) of Gujarat, voicing the demands of the people in 19 villages who would be affected by submersion due to the construction of dam. There were also views floated in favour of development but with a human face. It was subscribed by most that people who would be affected by environmental displacement should be accommodated well within the ambit of governance. There should be sufficient human intervention in order to accommodate the grievances of people who sacrifice their livelihood, dwelling, fodder and cultural roots in the name of development.

In many cases the government offered cash compensation. Since there is no sufficient mechanism to verify accountability, the people often complain of rampant corruption in the entire process of delivering the payment. The bureaucratic mannerisms are anti-poor and leave people with no option but to accept whatever compensation offered to them. Those people who used to rely on cultivable land now lost their only source of income. There is no guarantee from the part of the authorities that the land which they get in return as compensation in some other parts are equally cultivable, whether there is irrigation facility or whether the same crop can be grown depending on the quality of the soil. Productivity of land is an important factor and any kind of future yield from the land they get depends on the fertility and utility of the land. People are not going to benefit from land which does not suit their requirements. A serious lack of guideline is there in this regard for the exchange land offered to them. The kind of land offered to them is tokenism and it seldom stands for fulfilling the basic criteria of the human condition basic for the sustainability of life. Those who for generations used to cultivate a particular type of grain or crop would be forced to reconsider their traditional occupation but prefer to migrate to urban areas.

Many people therefore prefer cash compensation to land compensation since it drains and demoralises them socially, economically and psychologically. This forced migration to new occupations further deteriorates their economic status as the cash compensation given to them is not sufficient enough to rebuild life again in a new location. People move from a state of livelihood and sustainability to homelessness and joblessness. This wrecks their motivation to passively submit to the developmental initiatives and instead they seek to mobilise themselves against this gruesome exploitation in the name of national development. The continuous agitation against the construction of big dams has brought about major revision of water projects and even call for a suspension of World Bank funding of large dams (Sethi, 1993).

Society and ecological movements

These two movements are emblematic of the social embedding of the ecological movements, especially seen from the vantage point of society. These are not only mere movements for livelihood of the poor but also ways to find their own voices lost in history. They have explored and found new ways to voice their resistance and thereby breaking their political silence against development. They have made a mark in the classical development debate by articulating their culturally shared community knowledge of the local ecosystem. Thereby they have been able to break the hegemony of knowledge from the above whose locus is the urban centric western model of development.

Political affirmation through ecological movements brought migrants a sense of agency, especially to those families and relatives who are tied to their land and community. In many instances, it was also a mode of political struggle against marginalisation and the existing power structure in society. They have transformed from a group of scattered unidentified masses to people who have a strong sense of the political structure and economic exploitation. Local leadership and political mobilisation in the name of protection of nature provided a language of critique against development and modernisation through their own culturally embedded themes and images. It reflected the ecological short sightedness of the governments and exposed the illusion of a laissez-faire notion of holistic development. It made people realise that it is important to build checks and balances against state power. It has redirected the development policy debates from the status of a hegemonic discourse towards the need for a locally evolved sustainable model of development which is inclusive as well as participatory.

Part III

Marginalised by development and the question of livelihood

Emerging sociological perspectives on environmental movements in India

The sociological implication of environmental issues is an emerging area of concern among sociologists. This has gained momentum only in the recent years. There are only a few studies in this field. Although this is a matter of global concern, the focus here is only on the social and economic aspects of the ecological destruction in Indian context. Activists, voluntary organisations, research institutions and environmental scientists have begun to study the environmental situation in India. Many reports on India's environment were already being published by various NGO's like the Centre for Science and Environment.

Issues that emerged as a result of forced migration imply the analysis of the following aspects:

1 To pay attention to the social impact of industrialisation especially through building of large dams and construction of factories.
2 To understand what it means to live in an ecological balance state in terms of community life and its impact on changes in value system.
3 To address the conflicting relationship between economic development and social development.
4 To discuss the social inequality caused by large-scale displacement.
5 To analyse the people-centred approach and environmental protection.
6 To discuss the issue of social injustice caused by rapid industrialisation.
7 To evaluate the effect of losing access to some of the basic amenities supplied by nature such as water, fuel, wild fruits, roots, medical plants, cattle stock, pastures etc.
8 To anticipate the future from the current pattern of the use of the natural resources (air, rivers, mineral deposits, water tables and forest), and its impact on poor tribals and peasant communities and Dalits.
9 Institutional approach as to how various social institutions such as culture, religion and social life are affected by large-scale displacement and migration.
10 To explore the caste, class and gender dimensions of environmental movement.
11 To evaluate the question of rights and justice in the context of large-scale human rights violations due to dam construction and massive water projects.
12 To assess the nature of conflict, struggle and protest involved in the environment protection.
13 To critically evaluate the environmental policies and its impact on people.
14 To study the nature and extent of land alienation in India due to displacement.
15 To observe the nature, extent and impact of civil society in the environmental movement.
16 To demonstrate the problems incurred in the political rationality of development.
17 To discuss the role of civil society in environmental movements.

Land alienation, struggles of migrants and
environmental movements in India

Traditionally the people of rural India are largely dependent directly on nature for their everyday basic necessities. Most of them depend on water sources in their village which is mostly a river or a well or a stream. So is the case with their daily food supply. They collect firewood from the nearby forest or wasteland. They also have the knowledge of useful medicinal plants available in the local area. So the neighbourhood is a concept which is unique to rural India. Due to displacement, people lose the support of their relatives and neighbours. People are connected with each other through various rituals such as birth, education, marriage and death. People draw the meaning of their life attributing meaning to certain spaces which are marked as sacred. Rivers, trees, mountains, pastures, cattle and seasons all form part of their cultural and religious life.

Now the poor peasants are being pulled out of their permanent settlements and thrown into barren land of uncertainty and misery. This has led to forced migration. This not only pushes the people to economic misery but also uproots them from their social life. Therefore, in order to ensure environmental sustainability it is imperative that the village life is preserved. Consumer market is increasingly bringing changes in the lifestyle and consumption pattern of the urban people and at the same time, large-scale evacuation of villages and migration to the urban centres in search of employment is also happening. Rapid urbanisation would redefine man's relation to land and nature from a benevolent provider to that of a commodity for utility.

In India, however, environmental movements 'arose out of the imperative of human survival. This was an environmentalism of the poor, which married the concern of social justice on the one hand with sustainability on the other. It argued that present patterns of resource use disadvantaged local communities and devastated the natural environment' (Guha 2013). According to Guha (2013), the history of environment movement in India is characterised by the environmentalism of the poor. The concern for land and forest has been seen as a concern of the peasants and tribals. It has not been analysed from the perspective of environmentalism. Though in the west the environmental movement started after the 1970s – only after the United Nations Conference on Human environment in Stockholm, 1972 – in India it began during the colonial period itself. The tension was mounting on the land alienation of the peasants and state appropriation of tribal land. One of the main objectives of colonialism in India is to extract the raw material from India and to boost the local market in Britain. Rapid industrialisation and capital formation of the British economy was possible only by exploitation of the natural resources of the native colonies. When the issue of taxation and commercialisation of agricultural crops became a serious issue, the use of non-cultivated land and forest did not come to the limelight.

Pouchepadass says, 'but students of peasant societies have long considered the forest as of peripheral importance, probably simply because it was situated at the periphery of the cultivated space' (1995: 2059). 'It has been too rarely examined from a truly ecological standpoint the effects of land colonisation, agricultural and animal husbandry practices, hunting and gathering by peasants, and the functioning and crises of agro systems (except in the special case of famine)' (1995: 2059). Thus the environmental movement in India can be traced back to the colonial period. The colonial intervention into the settlements of tribal and peasant community was met with strong resistance (Gadgil and Guha 1992). The conflicts in different regions were over the extraction of forest produce and encroachment over forest land and the loss of their customary claim over the land.

It is only recently that separate attention is paid to the environmental condition and analyse against the rural and tribal agitations. So far the trend has been to treat the struggle of the marginal sections as class struggle for resources and land ownership. Because of that 'systematic analytical research-based monographs on environmental movements are few and far-between. More often than not the struggles of the people on the issues of their livelihood and access to forest and other natural resources are coined as "environmental movements" ' (Shah 2004). The social distinction between agricultural land and forest land made only the latter as an environmental issue. In the agrarian history of British India, the cultivated land was seen more in terms of the agrarian relations around land and conflict over unequal distribution rather than the relation between nature and society.

Gadgil and Guha (1992) consider that the ecological context of agriculture such as fishing, forests, grazing land and irrigation has been conceptually delinked from the analysis on agrarian history. It has also been observed especially in anthropological and sociological studies that these sections were mostly referred to as a brief description of the flora and fauna of the region under study suggesting that they are non-existent part of social interaction and hold any importance in determining the social hierarchy. The human perception of the use of the nature changes and it also affects the way we use the environment.

The current liberalisation would also only increase the developmental gap between the rural and the urban areas (Guha 2013).

Role of civil society and migrants in environmental movements in India

Civil Society as a separate sphere of activity and involvement in nation building includes NGOs, Universities, Media and New Media and all those social agencies which are part of neither the state nor the market. In the recent years the middle class in India has become active and vocal in raising numerous environmental issues. The support base of many environmental movements has increased with the wide national and international support extended to many movements. Many activist groups and protest movements have been able to sustain its struggle extended for many years mostly because of the participation and persistent support of the civil society.

However, one of the important features of the civil society in India is its middle class nature. Once the movement takes a momentum and grow beyond the local environment it is the professionals who become more prominent and gain prime media attention (Harsh Sethi 1993). The main issue of contention then is how far the movement could sustain the original issues which were the backbone of the movement. There is a tendency to move away from the core issues to which the movement has originally rendered its spirit.

At this juncture it is important to note that many migrants are now more aware of the urban state system and the functional way of employing things. They now join hands with the NGO activists to preserve their land from the onslaught of market economy. Another important aspect is the civil society participation and its nature of representation itself. The professionalism and the urban nature of articulation of the issue may not be always in consistent with the real method of evaluating the nature and extent of the problem faced by the victims and their family. Many times when the issue becomes more on compensation for the acquisition of land by the state the negotiation is reduced to the actual market value of the land only; not on the real trauma of displacement and uprootedness from their traditional ancestral land. The disruption of the social and community life does not become a concern of the professional middle-class activists and supporters who are mostly from the urban areas. Those who are left in the middle are these migrants who had hope in the middle class to support their cause.

As the movement grows and achieves national and international attention, the primary question one has to ask is whose rights are to be protected? The power dynamic within the movement is hard to ignore. The movement faces threat from the state and market forces and the constant torture from the police force the migrant families to make negotiations at different levels even on the main goal of the struggle itself. Then the question remains as to what is primary in a struggle: whether the rights of the victim, the survival of the movement or the larger mainstream support from civil society. Harsh Sethi argues: 'Almost invariably, in any coalition between the affected people and their middle class spokespersons – with the shift in power locus – issues tend to get clouded. While the journalists look for good copy, lawyers for the vital legal points that they will argue in court, and film-makers for the audience, we can easily forget that a grim battle for survival is taking place at the ground' (Harsh Sethi 1993: 139). The development of a middle-class orientation to the movement has therefore its own limitation and advantages for these migrant populations who have lost almost everything.

Amita Baviskar's (1995) study of the nature of involvement of NBA in the adivasi struggle in Madhya Pradesh further engages with this growing tension in the environmental movement. The caste (rich Patidars) and tribal communities (adivasi Bhilalas), though had conflict between each other, have come together under the banner of NBA against the drive to construct the dam in the region. The tribal resistance has been largely under the leadership of the activists of NBA. The main focus of NBA has been to voice the issue of development. Therefore, their nature of involvement is such that it tries to bring to the fore the anti-development policy of the state. They try to raise the consciousness of the people against the state apathy to the cause of the people. Their course of action is to build up networks and linkages from local to national level and campaign for a movement which would enable the people to have their voices heard by the public. Since the city activists have a greater part to play in the decision-making process their relation with the local adivasi is far from satisfactory. The real representation of the tribals is not possible unless they have equal participation in the middle class–oriented decision-making process.

At the later stage, many become regular migrants looking for seasonal employment in urban areas. There was constant migration despite the claim of the consolidation of the movement. Many started dissolutioned with the movement and started accepting the government terms and conditions.

Nature and its varied resources is a direct source of livelihood for most of the migrants for whom modern urban employment opportunities are hardly an alternative. At one level migrants are gradually shifting to modern economy and also adapting to modern urban settings as they can no longer sustain themselves solely based on traditional mode of sustainable livelihood. Nevertheless, their cord with nature continues to remain strong and the love for nature and protection and preservation comes naturally to them. There is a strong sense of being alienated from their cultural roots. Hence, the struggle of these forced migrants has compound facets which need to be articulated from the perspective of the migrants themselves.

Conclusion

By the end of the last century we heard the voices of the forced migrants as 'others' who have hitherto remained silent observers of the process of development. Their mobilisation and collective action towards finding a voice for themselves and finding out new means of political articulation became the salient feature of our contemporary political process. These 'others' are those whose voices are hitherto hidden from the mainstream, and the present-day development scenario had delimited their natural link with ecosystem. Among these migrant population includes the tribals, Dalits, women and the rural population who are the most affected group of the development projects. Followed by the impact of the forces of economic growth, the sustenance of their ecosystem has become a difficult proposition. Traditional ecosystem is now in the hands of development planners and in the easy reach of the corporate interest. The effort to build up an economic system which facilitates economic growth underplayed the need to directly engage with people whose livelihood is out of their reach now (Savyasaachi 2012). Therefore, they have the legitimate claim to voice and articulate the effect of the political rationality of current development projects, especially as their migrant ecology is at stake.

In order to analyse the modern means of production and distribution of commodity it becomes essential to look beyond the accepted norms of economy. The current economic system disseminates values that are directly growth intensive and indirectly influence our notion of livelihood. Traditional agrarian fields of yesteryears are gradually changing and propel the need to reconsider the notion of livelihood. Locally embedded labour force has no space in the neo-liberal labour market. People are in constant search for employment migrating to urban areas under a system which offers an urban ecological system which has its uniting force in the neo-liberal economic process of production and reproduction. Labour becomes under the influence of the universal process of production and distribution and therefore encounters a serious setback, and forces to move in the direction away from socially embedded working conditions.

Forced migrants now realise that a whole chain of hidden structure of intervention from multiple economic domains curbs their nurtured relation with their soil. Set under the condition of new developmental interventions it redefines their sense of belonging to the ecosystem. The power of the authorities who are part of the chain of the capitalist mode of relations of production needs to be understood not in isolation but as an overarching system which has a universal impact on our global ecosystem. Alternative modes of action which bring forth new networks become the immediate need of the hour. The voices from the unknown migrant 'other' from the below then have to become the voices of all.

It is the poor and the marginalised sections who were historically the real champions of environment protection and preservation. Ultimately the modern environment struggle is centred on the issue of livelihood. Any kind of failure in access and displacement out of these basic material resources further intensify their pauperisation which in turn increase the level of rural poverty. The resistance of people against developmental policies should be seen as resistance of these forced migrants in the present and the future migrants.

Hence to conclude I suggest that migrant ecology is the future. It is man as migrant theoretically which brings clarity to things. The future of Indians is migration. It has to be protected by considering migrant as a part of social ecology. However, in today's world migrant is subservient to the system where displacement is seen as wisdom of urban ecology. Instead I propose migrant ecology as a perspective to deal with things.

Notes

1 Bookchin, Murray. What Is Social Ecology? http://dwardmac.pitzer.edu/Anarchist_Archives/bookchin/socecol.html.
2 Ibid.
3 Latour, Bruno. (1993). *We Have Never Been Modern*. Cambridge: Harvard University Press.
4 Bookchin, What Is Social Ecology?
5 Bookchin, Murray. (1982). *The Ecology of Freedom: The Emergence and Dissolution of Hierarchy*. California: Cheshire Books.
6 Ibid. p. 323.
7 Ibid.
8 Hanns, Peter. (2006). Vitalizing Nature in the Enlightenment. H-HistGeog, H-Net Reviews. January.www.h-net.org/reviews/showrev.php?id=11323

9 Bookchin, *The Ecology of Freedom.*
10 Ibid.
11 Mies, M. and V. Shiva. (1993). *Ecofeminism.* London: Zed Books; Mitra, A. (1993a). "Chipko: An Unfinished Mission," *Down to Earth.* 30 April: 25–51; also see Mitra, A. (1993b). "There Can Be No Development without Women: Interview with Gayatri Devi," *Down to Earth.* 30 April: 50–51.
12 Jain, Shobitha, Standing Up for Trees: Women's Role in Chipko Movement. www.fao.org/docrep/ro465e/ro465e03.htm. Accessed on 13 December 2015.
13 Government of India. (1985). *National Forest Policy.* New Delhi: Ministry of Environment; Government of India. (1986). *The Environment (Protection) Act, 1986. (Act No. 29 0/1986).* New Delhi: Ministry of Law and Justice.
14 Rangan, Haripriya. (2004). "From Chipko to Uttaranchal: The Environment of Protest and Development in the Indian Himalaya," in Richard Peet and Michael Watts (eds). *Liberation Ecologies: Environment, Development, Social Movements.* London: Routledge.
15 Dwivedi, Ranjit (1998). "Resisting Dams and Development: Contemporary Significance of the Campaign against the Narmada Projects in India," *European Journal of Development Research.* 10(2):135–183.

References

Baviskar, Amita. (1995). *In the Belly of the River: Tribal Conflicts over Development in the Narmada Valley.* Delhi: Oxford University Press.

Gadgil, Madhav and Ramachandra Guha. (1992). *This Fissured Land: An Ecological History of India.* Berkeley: University of California Press.

Guha, Ramachandra. (ed.). (1994). *Social Ecology.* Delhi: Oxford University Press.

Pouchepadass, Jacques. (1995). "Colonialism and Environment in India: Comparative Perspective," *Economic and Political Weekly*, 30(33), August 19: 2059–2067.

Savyasaachi. (2012). "Struggles for Adivasi Livelihoods: Reclaiming the Foundational Value of Work," *Economic and Political Weekly*, 47(31), August 4: 27–31.

Sethi, Harsh. (1993). "Survival of Democracy: Ecological Struggles in India," in Ponna Wignaraja (ed.). *New Social Movements in the South, Empowering the People.* New Delhi: Vistaar Publications. pp. 20–32.

Shah, Ghanashyam. (2004). *Social Movements in India: A Review of Literature.* New Delhi: Sage Publications.

14 Spaces of recognition of climate migrants in India

Question of rights and responsibilities

Bratati Dey

Amna Khatun, a 21-year-old girl has completed her under graduate course from a college in the Sundarban region in West Bengal. According to her version, she has lot of capability to fulfil her dream courier. Unfortunately due to the effect of Aila (Tropical cyclone), her family displaced from their known place. Due to family pressure, she got married. Her family has started new livelihood practices in a new place. It is not a fairy tale of their life. Deepak Mondal is a 42-year-old young person. Present occupation is a labourer in a boat near Bagbazar Ghat in Kolkata metropolitan city. He had been displaced from his homeland Baripoda in Orissa due to the vast effect of super cyclones. A widespread view that is gaining ground is that climate-related migration could evolve into a global crisis by displacing a large number of people from their homes and forcing them to flee. In particular, it seems likely that significant numbers of people will be displaced either temporarily or permanently from their homes as a consequence of global warming (Stern 2007). From that point of view, this paper has some significant research questions. Available scientific evidence indicates that a large number of people might be displaced due to climate change but there is no internationally agreed definition of the term 'climate refugee' and the extent to which these displaced persons constitute a separate identifiable group. Without recognition these people have no citizen rights to enjoy their life. Predictions show that livelihood sources of the poorest will be diminished, therefore in order to cope with the shocks many will be migrating to urban centres. Holding Lefebvre's (1991) theory that space is created through manipulation, negotiation and appropriation, this paper argues that climate migrants in India are creating their own space in this global city as they are pushed in and forced to survive. This paper tries to emphasise on the relation between space and migration. Where spaces are continuously changing by groups of people and their practice on it, in that point of view, spaces are recreated by perception of people's habit and culture. Migrant people have their own culture which is superimposed on another existing space that creates new form of social problem. This paper focuses on climate migrants as a social group but recognition of new spaces creates more trouble for livelihood pattern. This paper also states the present scenario of climate migrants in India and examines the state of citizenship for the migrants as climate migration in the age of globalisation switches between the 'national citizen' and the changing notion of increased diversities. In this context, this paper focuses on questions of social rights and responsibilities. As a social scientist and being a geographer, I try to focus on who are symbolised as climate migrants. There is no concrete definition of climate migrant as well as no transparent policy for them. To recognise a new citizenship in space a new development model needs to be discussed with the help of postmodern space theory. Regarding the above discussion, this paper tries to search social justice for them.

Debates over the relation between climate change and migration often take place with reference to debates that have arisen over the last two decades in the context of environmental refugees or environmental migrants (Kniventon et al. 2008).[1] The concept of environmental refugees was first introduced by Lester Brown of the World Watch Institute in the 1970s (Black 2001). The concept became popular after studies by El-Hinnawi (1985) and Jacobson (1988) on forced migration of people due to environmental degradation and natural disasters. Myers and Kent (1995) and Norman are works on environmental refugees. Myers believes that the issue of environmental refugees 'promises to rank as one of the foremost human crisis of our times' (Myers 1997). Myers claims that there were at least 25 million environmental refugees in the mid-1990s. Again, he argues that when global warming takes hold, there could be as many as 200 million people displaced by disruptions of the monsoon system and other rainfall regimes by droughts of unprecedented severity and duration and by sea level rise and coastal flooding (Myers 2005). The debate over climate migrants has been often criticised on the ground that there is no accepted definition of climate migrants. Without an agreed definition, it is very difficult to say who can be categorised as climate migrants. Several researchers cast serious doubts on the predictions on numbers of climate migrants. The evidence put forward so far to link environmental factors to forced migration and refugees is often not scientifically or factually rigorous (Renaud et al. 2007).

Concepts of climate refugees or climate migrants

In 1990, the IPCC noted that the greatest single impact of climate change could be on human migration – with millions of people displaced by shoreline erosion, coastal flooding and agricultural disruption. Since then, various analysts have tried to put numbers on future flows of climate migrants (sometimes called 'climate refugees') – the most widely repeated prediction being 200 million by 2050.

The meteorological impact of climate change can be divided into two distinct drivers of migration: climate processes such as sea level rise, salinisation of agricultural land, desertification and growing water scarcity; and climate events such as flooding, storms and glacial lake outburst floods. But non-climate drivers, such as government policy, population growth and community-level resilience to natural disaster, are also important. All contribute to the degree of vulnerability people experience. The problem is one of time (the speed of change) and scale (the number of people it will affect).[2]

One immediately contentious issue is whether people displaced by climate change should be defined as 'climate refugees' or as 'climate migrants'. However, the use of the word *refugee* to describe those fleeing from environmental pressures is not strictly accurate under international law. The United Nations' 1951 Convention and 1967 Protocol relating to the status of refugees are clear that the term should be restricted to those fleeing persecution: 'A refugee is a person who, owing to a well-founded fear of being persecuted for reasons of race, religion, nationality, membership of a particular social group, or political opinion, is outside the country of his nationality, and is unable to or, owing to such fear, is unwilling to avail himself of the protection of that country'.[3]

If the term 'climate refugee' is problematic it is still used, in part, for lack of a good alternative. 'Climate migrant' implies the 'pull' of the destination more than the 'push' of the source country and carries negative connotations which reduce the implied responsibility of the international community for their welfare. But for lack of an adequate definition under international law, such migrants are almost invisible to the international system.

The International Organization for Migration (IOM) proposes the following definition: 'Environmental migrants are persons or groups of persons, who, for compelling reasons of sudden or progressive changes in the environment that adversely affect their lives or living conditions, are obliged to leave their habitual homes, or chose to do so, either temporarily or permanently, and who move either within their country or abroad.'

Climate migration is not a new phenomenon. Archaeological evidence suggests that human settlement patterns have responded repeatedly to changes in the climate. Migration is (and always has been) an important mechanism to deal with climate stress. Pastoralist societies have of course habitually migrated, with their animals, from water source to grazing lands in response to drought as well as part of their normal mode of life. When climate stresses coincide with economic or social stresses, the potential for forced migration from rural areas increases significantly.

Driving displacement has occurred in three different ways. 'First, warming of the atmosphere in some regions will reduce the agricultural potential and undermine the ecosystem services such as fertile soil and water affecting people's livelihoods. Second, increasing extreme weather events will generate mass displacement. Third, sea level rise will destroy the low-lying coastal areas and millions of people who will have to relocate permanently.[4]

Indian scenario

India is a climatologically diverse country. Its climate is characteristic of the subregional physical features, which also typifies its very diverse biological reserves and natural resources. Climatologically, the entire Indian region is divided as western Himalayas, north-west, north-east, northern central region, eastern coast, western coast and the interior plateau.

Climatologically India is at risk of accelerated sea level rise, intensification of cyclones, drought, heavy rainfall and large storm surges.

In India two types of displacement occurred due to climate change. First, increased migration is likely within India due to the effects of climate change such as drought, desertification, sea level rise, water scarcity and low food productivity. Second, climate change might lead to increased flow of migrants from neighbouring countries due to the accelerated effects of climate change.

As the variation of regional pattern in India, migration pattern has been diversified. For example, in western part of India climate change is expected to increase the severity of drought. Climate change is expected to increase drought in semi-arid peninsular India and western India, leading to further immiserisation of the landless and small and marginal farmers, who are typically forced to migrate more often to cities (Revi 2008).

A large part of the Himalayan valley regions of India are at risk of heavy rainfall. For example, prolonged heavy down pour on 16 and 17 June 2013 resembled 'cloud burst' (except for amount of precipitation of 100 mm/h) type event in the Kedarnath valley and surrounding areas that damaged the banks of River Mandakini for 18 km between Kedarnath and Sonprayag, and completely washed away Gaurikund (1,990 masl), Rambara (2,740 masl) and Kedarnath (3,546 masl) towns. The roads and footpath between Gaurikund and Kedarnath were also damaged. A large number of people migrate from their known place to nearby cities for survival.

India is a country with the second largest population having a large population located in the low elevation coastal zone. The Indian region is densely populated, stretches over 7,500 km, and is inhabited by more than a 100 million people in nine coastal states (McGranahan et al. 2007). Recent observation suggests that the sea level has risen 2.5 mm per year since the 1950s along the Indian coast. A one-metre sea level rise is projected to displace approximately 7.1 million people in India and about 5,764 sq. km of land area will be lost. Major areas of the mega cities of India such as Mumbai and Kolkata are at risk of sea level rise and storm surges which may induce people to migrate from the areas near to the sea. The Telegraph (2006) reported that submergence of the Lohachara Island in India's Sundarban has led people to move to the nearby Sagar Island. Climate change might lead to increased flow of migrants from neighbouring countries. As many as 120 million people could be rendered homeless by 2100 both in India and in Bangladesh due to sea level rise and given the proximity of Bangladesh to India much of the people will end up as migrants in Indian cities which are already facing resource scarcity (Rajan 2008).

Cyclone and the Sundarban

Out of 102 islands in the Indian Sundarbans, about 54 are inhabited and the rest are notified as reserved forest. People living on these islands are mostly migrants from other parts of West Bengal or Bangladesh. The islands lying further South (on the margins of the forest) and closer to the Bangladesh border have migrants mostly from Bangladesh, with immigrants still crossing the border and settling into the Sundarbans. These islands on the southern fringes are part of the active delta, being constantly configured and reconfigured by tidal movements in the rivers. The areas further up and nearer to Kolkata are parts of the stable delta. In the stable delta or in areas which are connected to the mainland of West Bengal, prevalent modes of transport are rickshaws, motor-driven three wheelers (often referred to as autos), buses and trains. However, the areas lying further south and surrounding the forests have mechanised boats (locally called bhatbhati) or non-motorised boats (dinghies) as the dominant mode of transport that connects otherwise isolated islands. Most of these islands have brick-paved roads which only allow cycle-vans (three-wheeled cycles with raised platform to carry goods and also people) to ply. These roads are few and vans ply as far as roads exist. Beyond roads are mud embankments or bunds, which serve as pathways connecting one part of an island to another. People's life on the southernmost islands revolves around land, water and forest. Although agriculture remains a source of livelihood for the islanders, the brackishness of rivers makes agriculture unsuitable and uncertain. Winter cultivation is virtually non-existent for want of fresh water. Poor families, especially those having very little or no land, rely on rivers for marine resources such as fish, prawn or crab. Forest is an important source of livelihood for poor families. The families frequent forests in search of firewood, wood, honey etc. People are engaged in livelihood activities that are physically demanding and challenging. The islands lying further south and on the margins of the forest are inhabited predominantly by people from Scheduled Caste (SC) communities. The Sundarbans also has a sizeable proportion of tribal population. About 42 per cent of the total population of the Sundarbans are from Scheduled Caste and Scheduled Tribe (ST) communities as against 28.5 per cent for the whole of West Bengal (see Tables 14.1 and 14.2 for Block-wise distribution of SCs, STs and other population). Among the 13 Sundarbans Blocks of the district of South 24 Parganas, Gosaba is one of the Southernmost Blocks, others being Basanti, Patharpratima, Kakdwip, Namkhana and Sagar. It is quite apparent from the people's livelihood options and the social composition of the population that people living in these Blocks of the

Table 14.1 Block-wise distribution of population of the Sundarbans in the district of North 24 Parganas

Blocks	Population	Male	Female	SC	ST	Sex ratio
Haroa	214,401	111,080	103,321	50,636	12,728	930
Minakhan	199,084	101,827	97,257	1,403	283	955
Sandeshkhali I	164,465	83,925	80,540	50,812	42,674	960
Sandeshkhali II	160,976	81,921	79,055	72,300	37,695	965
Hasnabad	203,262	104,019	99,243	51,295	7,492	954
Hingalganj	174,545	88,937	85,608	115,227	12,743	963

Source: Census of India 2011, Primary Census Abstract (www.censusindia.gov.in).

Table 14.2 Block-wise distribution of population of the Sundarbans in the district of South 24 Parganas

Blocks	Population	Male	Female	SC	ST	Sex ratio
Canning I	304,724	155,126	149,598	144,906	3,710	964
Canning II	252,523	128,438	124,055	52,859	14,910	965
Mathurapur I	195,104	100,093	95,011	58,636	496	949
Mathurapur II	220,839	113,831	107,008	52,342	4,643	940
Joynagar I	263,151	134,996	128,185	102,645	80	949
Joynagar II	252,164	128,858	123,306	85,587	1,046	956
Kultali	229,053	117,562	111,491	104,193	9,672	948
Basanti	336,717	171,279	165,438	119,631	20,060	965
Gosaba	246,598	125,910	120,688	154,584	23,343	958
Kakdwip	281,963	144,120	137,843	97,944	1,836	956
Sagar	212,037	109,468	102,569	56,261	854	936
Namkhana	182,830	93,351	89,429	47,260	741	957
Patharpratima	331,823	169,422	162,401	76,163	2,640	

Source: Census of India 2011, Primary Census Abstract (www.censusindia.gov.in).

Sundarbans remain socially and economically neglected.[5] Following table has been given some social information regarding this area.

The above-mentioned tables have focused on demographic pattern of society in this area. This area is densely populated, and sex ratio pattern is in favourable condition. In Sundarban area Schedule Caste population are majority and they have migrated from Bangladesh or near district from Medinipur.[6] According to geographical location and physical pattern of this area, it is not fertile for agriculture. Other sources of economic activity also face difficult problem due to physical setup. People have developed their own livelihood survival method for adaptation.

Natural disaster struck the District of North 24 Parganas – a chronological data

1983: Tornado at Gaighata Development Block.

1988: Very severe Cyclone at Hingalganj Development Block.

1995: Breech of embankment on 15 and 16 May in Basirhat Sub-Division. 1,077 houses were fully and 496 houses were partly damaged.

1996: Heavy rainfall occurred on 27 and 28 October with consequent breach of embankment in Basirhat Sub-Division. 48 houses were fully damaged.

1998: Flood in Bongaon Sub-Division due to incessant rains and overflow of river Ichhamati. Dates of occurrence were 14.08.1998 and 15.08.1998. Total 60.30 sq. km area and 96.700 people were affected. 10,059 houses were affected, out of which 3,457 were fully damaged. 18,483 persons were rescued and sheltered in 112 relief camps. Cyclone of high magnitude severely affected 466 sq. km of area in 10 blocks and 2 municipalities. Number of cattle lost was 773. And 2,16,622 people were affected by the calamity.

1999: The district was affected by drought during May–June 1999. Rainfall up to April, 1999 was only 6.5 mm. In the district 7 lakh people, 137,887 livestock were affected and from 4 per cent to 40 per cent Boro paddy was lost.

Due to heavy rainfall from 21 to 24.09.1999, a vast area of 14 blocks and 12 municipalities in all the 5 sub-divisions of the district were water-logged. And 4,50,000 people were affected.

2000: Unprecedented incessant heavy shower from 17.09.2000 to 21.09.2000 followed by very high discharge from the dams and barrages caused huge flood all over the state having highest intensity in about last two centuries. The water level of Churni rose to an unprecedented 11 m against highest recorded level of 9.798 m in 1978. A vast area of Bongaon Sub-Division and part of Barasat (Sadar) and Basirhat Sub-Divisions were inundated; some places for more than a fortnight. The year 2000 may be designated as the year of the worst precipitation in terms of quantum, intensity and duration. More than 960 sq. km areas were affected in the district causing damages to houses, crops, school buildings etc. More than 19 lakh people were affected and 2, 65.000 houses were destroyed or damaged. 652 mouzas were declared as flood affected. Total financial loss was Rs 62,207 lakh.

2001: There was a severe thunder storm on 20.05.2001 at Gaighata Block under Bongaon Sub-Division. The most affected GPs were Jaleswar II. Duma and Dharampur I. Four human lives were lost.

2002: On 25.05.2002, embankment of Bermajur Gram Panchayat on river Choto Kalagachi under Sandeshkhali II Block breached affecting 2,500 people. A severe cyclonic storm passed over Sandeshkhali I, Sandeshkhali II, Hingalgunj and Minakhan blocks on 11.11.2002 affecting 5,483 people. 195 houses were destroyed and 731 were damaged due to the calamity. The event of crop area damaged was 644.30 ha, with monetary loss of Rs 556,82,000. There was a very severe cyclonic storm on 03.04.2002 at 3 AM in Bongaon Sub-Division. About 65,000 people were affected. Three human lives were lost in Bongaon Block and five goats and 1,025 poultry were lost. The extent of damage caused to power and housing sector was Rs 8.5 lakh and 253 lakh, respectively.

2003: Consequent upon the development of low pressure in the Bihar Plateau Region, the South Bengal districts including North 24 Parganas experienced heavy rainfall during the period from 16.10.2003 to 10.10.2003. Breach of embankments and erosion were reported from Hingalgunj, Sandesh-khali I and Sandeshkhali I blocks. Due to the calamity, 14 blocks in North 24 Parganas district were severely affected. 510,958 people were affected and 8,539 houses were fully and 15,131 were partly damaged. Losses in crop sector and fisheries sector were Rs 282.35 lakh and 2303.28 lakh, respectively.

2004: Due to incessant rain during the period from 10.09.2004 to 18.09.2004 a vast area of Bongaon, Basirhat, Barrackpore and Barasat (Sadar) Sub-Divisions was waterlogged. 8,89,283 people were affected. Agricultural loss was estimated to Rs 6802.665 lakh. 359 mouzas were declared as flood affected. 18,384 dwelling houses were destroyed and 56,865 were damaged in urban and municipal areas. There was a violent storm on 06.10.2004 at 12 noon over paruipara, Biswaspara and Angrail within the jurisdiction of Jhowdanda Gram Panchayat under Gaighata Block. Two persons were seriously injured and admitted to hospital. And 29 houses were destroyed.

2006: Due to incessant rainfall a vast area of Basirhat, Bongaon, and part of Barasat and Barrackpore Sub-Divisions were affected. Eight persons died due to the natural calamity during the year 2006.

2007: This district was affected in two phases due to incessant rainfall in 2007. Average rainfall was 1,380 mm. Many people were affected.

2008: During the year 2008 this district was affected on different occasions by incessant rain with storm (16.09.2008, 26.09.2008, 25.10.2008). About 74,000 people were affected during the above calamities.

2009: On 25 May, Cyclone Aila hit coastal Bengal with a maximum wind speed of 120 kmph affecting over 1.5 million people. It swept across South Bengal particularly the deltaic Sundarbans killing people, their livestock and rendering thousands home-

less. Those living on the margins once again became marginalised. Many people having lost their land, houses and also
their family members. The saline water that broke through
embankments, flooded the villages, destroyed mud houses and
polluted rice fields.[7]

Hamlets have been reduced to wasteland – with submerged crops, up-
rooted trees, shattered homesteads and emaciated [and dead] cattle
all around. Ponds which have been the only source of portable water
lay contaminated and stinking. Not even stray dogs that survived the
disaster would go near them (Mukhopadhya 2009: 33).

Over 5.1 million people have been affected in 16 districts of West
Bengal. The damage impact assessment carried out by the Government
of West Bengal and the United Nations Development Programme
(UNDP) reported considerable deaths, of which 25 were caused by
a landslide in Darjeeling. Over 500,000 houses were damaged either
fully or partially. The storm was especially devastating for farmers
who were preparing to harvest rice and other crops. According to
media sources, the Sundarbans national reserve forest was worst-hit,
as many as three million people lived in the forests.

The damages due to Aila cyclone in Basirhat Sub-Division
(North 24 Parganas District, West Bengal) only

Blocks affected – 10 (all)
Municipalities affected – 3 (all)
Villages affected – 857
People affected – 677,662
Crop area damaged – 16,210 ha
Value of crop damaged – 42.49 crore
House damaged/destroyed – Fully 90,748, Partly 48,315
Embankments damaged – 45.5 km
Panchayat property damaged – 15.55 crore
Road damaged – 111 km (Value 11.1 crore)
          (Source: West Bengal Disaster Management Report 2013)

The social dimension of vulnerability

The mediating function of social factors in the relationship between
climate change and migration points to the fact that people do not
have access to the same resources when it comes to reacting or adapting
to environmental change. Vulnerability is therefore shaped by a wide

range of social variables that determine people's exposure to climate change. From a social sciences perspective, this would seem to go without saying; yet, studies on the climate change–migration nexus have long privileged top-down approaches in which so called hotspots are identified and mechanically understood as places where migration will occur – regardless of 'from below' considerations on the ways in which people will react and adapt. This is manifest in many of the available maps on the topic, in which one can see the geographical zones likely to be affected by climate change – but which say nothing of the social context. This includes for example gender, as changes in livelihood patterns affect men and women differently, not only because of their different social positions, but also because gender is known to influence the perception of risks (which is a crucial variable in migration strategies), as well as the way people experience displacement. Another core variable in the construction of vulnerability is of course class resources and wealth. Climate change affects disproportionally poor agrarian communities, precisely those that have the least resources to leave their home. The consequences of climate change thus vary according to the context, as the same environmental factor will have different impacts according to the characteristics of the people it affects. It follows that environmental degradation does not mechanically lead to displacement and that one should resist the 'tendency to equate populations at risk with population displacement' (Hugo 2008: 31).

Spaces of recognition

Above discussion shows that livelihood sources of the poorest will be diminished, therefore in order to cope with the shocks, many will be migrating to urban centres. Holding Lefebvre's theory that space is created through manipulation, negotiation and appropriation, this paper argues that climate migrants in India are creating their own space in this global city as they are pushed in and forced to survive. In his pioneer book *The Production of Space* (1991) he has introduced the concepts of the right to the city and the production of social space. This work has deeply influenced current urban theory, mainly within human geography, as seen in the current work of authors such as David Harvey, Dolores Hayden and Edward Soja, and in the contemporary discussions around the notion of spatial justice.

Soja introduces his epistemological approach to space. Three main concerns unfold from Soja's project. First and foremost, Soja makes the point that space is never given. It is never an 'empty box' to be filled, never only an entity. It is part of the general cultural web and like any cultural stage, it is not a mere background. On the contrary space is always a culturally constructed entity space constantly formed, changed, accepted or rejected. This is, however, a point that is made by many theorists on space. The most important contribution of Soja to the postmodernism's way of thinking is that he visualised the other way of looking at being and spatiality. He introduces the conception of 'Third Space'. Soja defines 'spatiality' as 'socially produced space'. Holding these concepts we have to focus on the life of climate migrants. Regarding above issues in Sundarbans area people have their own culture and livelihood pattern for survival. They have practised their own culture forming their social space. After migration, they have superimposed on new spaces. Soja introduced that this space might create third spaces. Migrant people are not capable to conceive new livelihood pattern so easily. This imaginative space creates trouble for their survival. More emphasis on this aspect is needed and policy makers should be sensitive about migrants' rights and responsibilities as citizens.

Protection of environmental migrants and states' responsibilities

As argued, the different terms referring to people who migrate in connection with environmental factors imply different representations of how states could or should treat these people and of the protection that they should receive. The starting point of this complex and sensitive issue is the current absence of standards in defining this protection; indeed, none of the concepts mentioned above have a legal definition – leading to an institutional and normative vacuum.

In the absence of specific norms, one could try to rely on existing instruments and explore how they relate to the issues relating to environmental migrants. In the case of people moving within their own country (which, as argued above, is the most frequent case), existing soft law instruments and notably the Guiding Principles on Internal Displacement do recognise some environmental factors (e.g. disasters) as a cause for displacement. But they suffer from implementation challenges, which are due to problems of definition and to the non-binding nature of the principles.

Possible policy orientations

What are the policies that have been elaborated to respond to environmentally induced migration? And what are the policy orientations that could be envisaged to address the challenges raised by the movement of people in a context of environmental change? Given the heterogeneity in the types of climate stress that can foster migration, it is worth distinguishing between different kinds of policy options.

First, there is the case of disasters and sudden climatic events. There have always been cyclones, floods or other natural catastrophes and most, if not all, regions of the world have experienced the challenge of addressing the situations of the persons concerned. The problem lies in the efficiency of the already existing mechanisms, especially if one assumes that climate change will increase the frequency and/or intensity of some kinds of disasters – thus putting humanitarian efforts under further stress. This calls for reinforcing rescue mechanisms and, in the case of less developed countries, for greater international solidarity, not least in making the necessary funds available. Overall, the main objective should therefore be to make a more extensive use of existing policy mechanisms and to adapt them to the specific challenges raised by climate change.

In some extreme cases, resettlement may constitute the appropriate policy, in order to enable large numbers of people to leave their home on a permanent basis. But these are not new policies either as resettlement has regularly been implemented in other contexts, especially in relation to large-scale infrastructure projects like dams. Again therefore, the relevant policy approach would be to improve existing policy options, through increased funding and international cooperation. This being said, resettlement is not an option for all the people concerned by progressive manifestations of climate change. There is therefore a need to envisage a much broader range of responses, to address the multifaceted challenges raised by slow environmental deterioration. At the local level, this could for example include measures to diversify economic activities in order to enable people to better adapt to climate change. This discussion highlights the fact that, even if environmental migration is regularly presented as a 'new' challenge requiring 'new' responses, there are actually a number of existing policy fields that can be relied upon to address the challenges it raises, including development strategy, humanitarian affairs, post-disaster interventions or immigration and admission policies. This is not to say that new normative or policy instruments are irrelevant; rather, it means that new instruments may not be a prior necessity to address the needs of the populations at risk and that an absence of consensus on the desirability of such new standards does not imply that nothing can be done.

- Planning for adaptation to climate change
- Foster adaptation alternatives that include migration

- Support disaster risk reduction and conflict mediation strategies
- Creating sustainable migration and development policies
- Developing new immigration policies
- Create alternative livelihoods where traditional livelihoods affected; imperative to reduce rural-to-urban migration (e.g. through relocation of small-scale industries to rural areas, environmentally sustainable shrimp farming etc.)

Conclusion

Climate change does have consequences in terms of human migration and mobility, and its impact can be expected to increase. But given the complexity of the relationship between environmental change and migration, it is worth recalling that climatic or natural hazards do not automatically lead to displacements.

The social dimension of vulnerability should be interpreted as an opportunity to increase people's ability to resist climate change. Indeed, if human beings were completely helpless in the face of nature and climate change, very little could be done. But they are not and this opens opportunities for local and international efforts in gathering knowledge, drafting measures and increasing protection. Provided that the necessary financial means are made available, even such an apparently unavoidable threat like rising sea levels could be partially counteracted. It also follows that, if environmental migration is fundamentally a political process, the actual number of people who will move cannot be predicted, but depends upon current and future efforts.

More knowledge is required to address the situation of people affected by environmental change and it is paramount to understand better the kind of patterns that develop out of it in order to envisage potentially successful policies. In addition, research on these issues requires increased cooperation between social and natural sciences, for instance, in the elaboration of complete and comparable databases. All in all, climate change is a process that exacerbates some of the most pressing issues of our time. It does not take place in a vacuum but is closely associated with underdevelopment, inequalities within and between countries, global justice and the lack of solidarity between states, human rights or human security. Climate change as a policy area may be relatively recent, but most of these issues represent long-standing challenges for states and the international community. It follows that policies that focus on the climate change–migration nexus must be accompanied by renewed efforts to combat the very context that make people vulnerable in the first place.

Notes

1 Panda, Architesh (2010), Climate Refugees: Implications for India, *Economic and Political Weekly*, Vol. XIV, No. 20, pp. 76–79.
2 IOM (2007), Discussion Note, "Migration and the Environment", International Organisation for Migration, Geneva, www.iom.int/jahia/webdav/shared/shared/mainsite/about_iom/en/council/94/MC_ NF_288.pdf
3 University of Oxford (2010), Centre on Migration, Policy and Society, Working Paper No. 79.
4 Panda, Climate Refugees, pp. 76–79.
5 Mukhopadhyay, Amitesh (2009), *Cyclone Aila and the Sundarbans: An Enquiry into the Disaster and Politics of Aid and Relief*, Mahanirban Calcutta Research Group, Kolkata.
6 Medinipur – District of West Bengal.
7 Mukhopadhyay, *Cyclone Aila and the Sundarbans*.

References

Black, Richard (2001), *Environmental Refugees: Myth or Reality?*, Working Paper No 34, New Issues in Refugee Research, UNHCR.

El-Hinnawi, E. (1985), *Environmental Refugees*, Nairobi: United Nations Environmental Programme.

Hugo, G. (2008), *Migration, Development and Environment*, Geneva: IOM International Organization for Migration.

IOM (2007), Discussion Note, Migration and the Environment, International Organisation for Migration, Geneva, www.iom.int/jahia/webdav/shared/shared/mainsite/about_iom/en/council/94/MC_NF_288.pdf

Jacobson, Jodi (1988), *Environmental Refugees: A Yardstick of Habitability*, World Watch Paper No 86, Washington, DC: World Watch Institute.

Kniveton, D., K. Schmidt Verkerk, C. Smith, and R. Black (2008), Climate Change and Migration: Improving Methodologies to Estimate Flows, IOM Migration Research Series, No. 33, International Organisation for Migration, University of Sussex, Brighton.

Lefebvre, Henri (1991), *The Production of Space*, USA: Wiley Blackwell.

McGranahan, G., D. Balk, and B. Anderson (2007), The Rising Tide: Assessing the Risks of Climate Change and Human Settlements in Low Elevation Coastal Zones, *Environment and Urbanisation*, 19(1): 17–37.

Mukhopadhyay, Amitesh (2009), *Cyclone Aila and the Sundarbans: An Enquiry into the Disaster and Politics of Aid and Relief*, Kolkata: Mahanirban Calcutta Research Group, Kolkata.

Myers, N. and J. Kent (1995), *Environmental Exodus: An Emergent Crisis in the Global Arena*, Washington, DC: The Climate Institute.

Myers, N. and J. Kent (2002), Environmental Refugees: A Growing Phenomenon of the 21st Century, *Philosophical Transactions of the Royal Society*, 357(1420): 609–613.

Myers, N. and J. Kent (2005), Environmental Refugees: An Emergent Security Issue, 13th Economic Forum, Prague, 23–27 May, www.osce.org/documents/eea/2005/05/14488_en.pdf, Accessed on 10 July 2009.

Panda, Architesh (2010), Climate Refugees: Implications for India, *Economic and Political Weekly*, 14, 20.

Rajan, Chella S. (2008), Blue Alert, Greenpeace India Society, www.green peaceindia.org

Renaud, F., J. J. Bogardi, O. Dun, and K. Warner (2007), Control, Adapt or Flee: How to Face Environmental Migration, InterSecTions, UNU-EHS, no. 5/2007, www.ehs.unu.edu/file.php?id=259

Revi, A. (2008), Climate Change Risk: An Adaptation and Mitigation Agenda for Indian Cities, *Environment and Urbanization*, 20(1): 207–229.

Stern, Nicholas (2007), *The Economics of Climate Change: The Stern Review*, Cambridge: Cambridge University Press.

Index